MATERIALS SCIENCE AND TECHNOLOGIES

AMINES GRAFTED CELLULOSE MATERIALS

MATERIALS SCIENCE AND TECHNOLOGIES

Additional books in this series can be found on Nova's website at:

https://www.novapublishers.com/catalog/index.php?cPath=23_29&seriesp=
Materials+Science+and+Technologies

Additional E-books in this series can be found on Nova's website at:

https://www.novapublishers.com/catalog/index.php?cPath=23_29&seriespe=
Materials+Science+and+Technologies

MATERIALS SCIENCE AND TECHNOLOGIES

AMINES GRAFTED CELLULOSE MATERIALS

NADEGE FOLLAIN

Nova Science Publishers, Inc.
New York

Copyright © 2010 by Nova Science Publishers, Inc.

All rights reserved. No part of this book may be reproduced, stored in a retrieval system or transmitted in any form or by any means: electronic, electrostatic, magnetic, tape, mechanical photocopying, recording or otherwise without the written permission of the Publisher.

For permission to use material from this book please contact us:
Telephone 631-231-7269; Fax 631-231-8175
Web Site: http://www.novapublishers.com

NOTICE TO THE READER

The Publisher has taken reasonable care in the preparation of this book, but makes no expressed or implied warranty of any kind and assumes no responsibility for any errors or omissions. No liability is assumed for incidental or consequential damages in connection with or arising out of information contained in this book. The Publisher shall not be liable for any special, consequential, or exemplary damages resulting, in whole or in part, from the readers' use of, or reliance upon, this material. Any parts of this book based on government reports are so indicated and copyright is claimed for those parts to the extent applicable to compilations of such works.

Independent verification should be sought for any data, advice or recommendations contained in this book. In addition, no responsibility is assumed by the publisher for any injury and/or damage to persons or property arising from any methods, products, instructions, ideas or otherwise contained in this publication.

This publication is designed to provide accurate and authoritative information with regard to the subject matter covered herein. It is sold with the clear understanding that the Publisher is not engaged in rendering legal or any other professional services. If legal or any other expert assistance is required, the services of a competent person should be sought. FROM A DECLARATION OF PARTICIPANTS JOINTLY ADOPTED BY A COMMITTEE OF THE AMERICAN BAR ASSOCIATION AND A COMMITTEE OF PUBLISHERS.

LIBRARY OF CONGRESS CATALOGING-IN-PUBLICATION DATA

ISBN : 978-1-61668-196-8

Available Upon Request

Published by Nova Science Publishers, Inc. ✝ New York

MATERIALS SCIENCE AND TECHNOLOGIES

AMINES GRAFTED CELLULOSE MATERIALS

NADEGE FOLLAIN

Nova Science Publishers, Inc().
New York

Copyright © 2010 by Nova Science Publishers, Inc.

All rights reserved. No part of this book may be reproduced, stored in a retrieval system or transmitted in any form or by any means: electronic, electrostatic, magnetic, tape, mechanical photocopying, recording or otherwise without the written permission of the Publisher.

For permission to use material from this book please contact us:
Telephone 631-231-7269; Fax 631-231-8175
Web Site: http://www.novapublishers.com

NOTICE TO THE READER

The Publisher has taken reasonable care in the preparation of this book, but makes no expressed or implied warranty of any kind and assumes no responsibility for any errors or omissions. No liability is assumed for incidental or consequential damages in connection with or arising out of information contained in this book. The Publisher shall not be liable for any special, consequential, or exemplary damages resulting, in whole or in part, from the readers' use of, or reliance upon, this material. Any parts of this book based on government reports are so indicated and copyright is claimed for those parts to the extent applicable to compilations of such works.

Independent verification should be sought for any data, advice or recommendations contained in this book. In addition, no responsibility is assumed by the publisher for any injury and/or damage to persons or property arising from any methods, products, instructions, ideas or otherwise contained in this publication.

This publication is designed to provide accurate and authoritative information with regard to the subject matter covered herein. It is sold with the clear understanding that the Publisher is not engaged in rendering legal or any other professional services. If legal or any other expert assistance is required, the services of a competent person should be sought. FROM A DECLARATION OF PARTICIPANTS JOINTLY ADOPTED BY A COMMITTEE OF THE AMERICAN BAR ASSOCIATION AND A COMMITTEE OF PUBLISHERS.

LIBRARY OF CONGRESS CATALOGING-IN-PUBLICATION DATA

ISBN : 978-1-61668-196-8

Available Upon Request

Published by Nova Science Publishers, Inc. + New York

Contents

Preface		**vii**
Chapter 1	Introduction	**1**
Chapter 2	Experimental Sections	**5**
Chapter 3	Results and Discussion	**15**
Conclusion		**55**
References		**57**
Index		**63**

PREFACE

*Nadege Follain**

Centre de Recherches sur les Macromolécules Végétales, (CERMAV-CNRS), affiliated with the Joseph Fourier University, Grenoble Cedex, France

Native cellulose is a structural material that is biosynthesized as microfibrils by a number of living organisms ranging from higher and lower plants, to some ameobae, sea animals, bacteria and fungi.-Depending on their origin, individual cellulose microfibrils have diameters from 2 to 20 nm, while their length can reach several tens of microns. The chemical modification of cellulose microfibrils is investigated for preparing new bio-based materials with end-use properties in the fields of adhesion, textile, detergent, paint, cosmetic, medicine, food, etc. Among the possible chemical modifications, the selective oxidation of the primary alcohol group of polysaccharides has been studied for more than a half-century. Recently, a method for selectively oxidizing primary alcohol groups of polysaccharides has been described in literature without degradation of products. The technique is based on a reaction catalyzed by 2,2,6,6-tetramethyl-1-piperidine oxoammonium radical (TEMPO) in presence of NaOBr, generated *in situ* by NaOCl and NaBr, the catalyst being regenerated during the reaction. The chemical modification is a way to modify and introduce specific functionalities leading to the development of new biopolymers in macromolecular prodrug carrier, bio-based composites, nanocomposites, for example. The polymer must be biodegradable and / or biocompatible and must contain appropriate functional sites for chemical conjugation. Despites on large interest on natural and synthetic biodegradable polymers

---

* Present Address of corresponding author: University of Rouen, Laboratory « Polymers, Biopolymers and Surfaces », UMR 6270 & FR3038 CNRS, 76821 Mont-Saint-Aignan cedex, France. Telephone: 33-235 14 66 98; Fax: 33-235 14 67 04; e-mail address: nadege.follain@univ-rouen.fr

investigated, important efforts are continuing to search for new systems, notably on cellulose.

In this report, the amidation of cellulose materials previously modified by carboxylation reaction is realized from the selection of some amines (cyclic and linear structures). The carboxylation is resulted from the TEMPO mediated oxidation of cellulose, leading to partially or totally oxidized cellulose, in presence of carbodiimide which is known to increase the reactivity of carboxyl groups toward amidation but used rarely for polysaccharides holding carboxyl moieties. Few reports were found in the literature on the use of carbodiimide in the preparation of cellulose conjugates with amines in order to develop new modified cellulose materials. The goals of this report are to develop ways of preparing the cellulose conjugates which can be water-soluble materials or water-insoluble materials, to identify linkage with carboxylated cellulose materials through amide bonds and to understand the obtained results following by FT-IR, NMR spectroscopies (at liquid and solid states) and electron paramagnetic resonance spectroscopy. The carboxyl content of oxidized cellulose materials after carboxylation and after amidation reactions are equally determined by titration curves of conductimetry and elemental analysis.

Keywords. Cellulose – Surface carboxylated cellulose nanocrystals – Oxidation – Amidation – EPR spectroscopy – Solid-state NMR Spectroscopy

Chapter 1

INTRODUCTION

Native cellulose, structural material biosynthesized, is produced in the form of microfibrils having diameters ranging from 2 to 20 nm depending on their origin, while their length can reach several tens of microns [1]. These microfibrils are made of perfectly aligned cellulose molecules organized into a nearly defect-free crystalline arrangement. Some imperfections, referred to as amorphous zones, along the microfibril length are hydrolyzed by acid treatment to produce nanocrystals called cellulose whiskers [2]. The shape and size of nanocrystals are more or less fixed by cellulose origin: different samples like tunicin, cotton, bacterial cellulose, parenchyma cell cellulose produced different sizes of nanocrystals even under similar hydrolysis conditions.

Recently, a method was found to catalyze the selective oxidation of primary hydroxyl groups in aqueous media [3-4]. This technique is based on a reaction catalyzed by 2,2,6,6-tetramethyl-1-piperidine oxoammonium radical (TEMPO) in presence of NaOBr, generated in situ by NaOCl and NaBr, the catalyst being regenerated during the reaction [3, 5-6]. This method was firstly proposed for water-soluble polysaccharides [5-9] such as starch, inulin, amylodextrin, pullulan, amylopectin, chitosan. The method has been later extended to water-insoluble polysaccharides such as cellulose, amylose and chitin [7-12]. In the case of cellulose I or native cellulose, the oxidation takes place only at the crystal surfaces which became negatively charged; a phenomenon that can be interesting for subsequent grafting or derivatization purposes [10]. Nevertheless, the cellulose can be fully oxidized to yield pure polyglucuronic acid only if regenerated or mercerized cellulose samples are used [9]. In the literature, the procedure leading to cellulose III allomorph from cellulose I is described from to an ammonia treatment [13]. This procedure known to increase the accessibility of crystalline cellulose [14] is realized at a solid-state that allows to maintain the integrity of the cellulose microfibrils while a substantial decrystallization combined to a reorganization of the intracrystalline hydrogen bond network of cellulose are achieved. By way of this procedure, the conversion of cellulose I into III_I improves its reactivity with respect to the TEMPO mediated oxidation system. If the cellulose I is used without any treatment as starting material, a partially conversion is realized due to its higher crystallinity and therefore the poor accessibility of primary hydroxyl groups. This approach introduced carboxyl groups at the surface of the crystals providing surface carboxylated cellulose nanocrystals materials. A number of applications have been described for carboxylated cellulose (also named oxidized cellulose) in the field of gelation, complexation, antifloculation activities, adhesion, biological activities and as well as a textile, paper, reinforced bio-based composites.

Polymers having polar groups like carboxylated cellulose are easily modified by esterification or amidation among others, the latter given generally better chemical resistance against hydrolysis. The establishment of an amide bond has been firstly proposed by Danishefsky [15] applied to a mucopolysaccharide. However, some reports claimed successful formation of amide link [15-16] whereas others reported failure in amidation [17-18]. Recently, the use of N-hydroxysuccinimide like an activating reagent was found to be effective for the EDAC mediated amidation [10, 18]. Water-soluble carbodiimides such as EDAC present the advantage of being soluble in aqueous media and can be eliminated by simple dialysis. The grafting conditions of amines on oxidized cellulose were extensively studied [10, 18-22] and the procedures are further and still based on the Bulpitt and Aeschlimann works [18]. This procedure is applied to water-soluble carboxylated cellulose and equally water-insoluble carboxylated cellulose. The selection of single terminally aminated molecules like grafted reagents corresponded to primary linear and cyclic amines.

An attempt to modify the cellulose material from totally oxidized cellulose (water-soluble cellulose) and partially oxidized cellulose (water-insoluble cellulose) samples is proposed to confer new features to expand the potential applications. The goals of this work are to describe a new way to get cellulose derivatives, to introduce the grafting mechanism, to investigate it from to a large range of characterization and to correlate a theoretical knowledge with the experimental results derived from conductimetry, elemental analysis and EPR, FTIR, NMR measurements. Considerable and promising efforts are going ahead to develop new macromolecular carrier systems and reinforcing structure in bio-based composites based on oxidized cellulose that is biocompatible and biodegradable polymer [23].

Chapter 2

EXPERIMENTAL SECTIONS

MATERIALS

Cellulose

A batch of cotton linters from Tubize Plastics (Rhodia - Belgium) was used as received.

Chemicals

For the oxidation reaction, 2,2,6,6-tetramethylpiperidine-1-oxyl radical (called TEMPO), sodium bromide and sodium hypochlorite from Aldrich were directly used.

Table 1. Grafting reagents

Grafting reagents	Molecular weight (g/mol)	Formula
EDAC	191.70	(ethylcarbodiimide structure) .HCl
NHS	115.09	(N-hydroxysuccinimide structure)

For the amidation reaction, N-(3-Dimethylaminopropyl)-N'-ethylcarbodiimide hydrochloride (EDAC) and N-hydroxysuccinimide (NHS) provided by Sigma Chemical (Table 1) were used as reagents.

For grafting, the amines were purchased from Aldrich: 4-amino 2,2,6,6-tetramethylpiperidine-1-oxyl radical (called 4-amino TEMPO), *n*-octylamine, *n*-butylamine and 2-methoxyethylamine (Table 2).

Table 2. Different amines used for grafting reaction.

Amines	Molecular weight (g/mol)	Formula
4-amino TEMPO	171.26	
n-octylamine	129.15	
n-butylamine	73.14	
2-methoxyethylamine	75	

Notations

1 eq of X represents the molar quantity of X for 1 mol of glycosyl unit (M = 162 g/mol).

PREPARATION OF CELLULOSE SUBSTRATES

Preparation of Cellulose III by Ammonia Treatment

In order to improve the accessibility of crystalline cellulose [24], the swelling of cellulose in ammonia or in molecules such as amines is a simple and classical way. This procedure leads to cellulose III_I from cellulose I (native cellulose) and has been frequently used to improve the reactivity of crystalline cellulose for the preparation of derivatives in better yields [25-26]. Indeed, this conversion carried out essentially at a solid-state process that keeps the integrity of the cellulose microfibrils while achieving a substantial decrystallization and a reorganization of the intra-crystalline hydrogen bond pattern of cellulose [27-28]. Ammonia or amines enter the cellulose crystals as guests, which not only distort the crystals but also modify the conformation of the hydroxymethyl group within the cellulose lattice itself [29-30]. The protocol to prepare cellulose III [13] involved the use of exploded gaseous ammonia (EG-NH_3) and corresponded to samples processed by Rhodia Acetow, following their patented process, which consists of treating cellulose samples with gaseous ammonia under high pressure followed by a rapid decompression [31-32]. This particular cellulose preparation was obtained from Rhodia, company which developed the technology for its use. The scientific basis for the "ammonia explosion" process has been achieved by Dale and collaborators [33-34].

HCl Hydrolysis

20 g of cotton linters were hydrolyzed with 1 L of 2.5 M HCl at 100°C for 20 min. The hydrolyzate was filtered and washed with water until neutral pH. The weight loss resulting from the hydrolysis step was around 10% for cotton linters. It is estimated that this product loss corresponds to the hydrolysis of the amorphous zones.

TEMPO-MEDIATED OXIDATION

Oxidation experiments were carried out on cellulose III_I and HCl-hydrolyzed cotton linters cellulose as previously published [12-13] with minor modifications [22]. In a typical run, cellulose samples (1.95 g, 12 mmol glycosyl units) were dispersed in distilled water (180 mL) for 3 minutes with a high speed T25 basic Ultra-Turax homogenizer (Ika-Labortechnik, Staufen, Germany). 90 mL of water used to wash the homogenizer was then added to the suspension. TEMPO (30 mg, 0.19 mmol), NaBr (0.63 g, 6.1 mmol) and NaOCl (1.76 M solution, 1.5 mL, 2.64 mmol) was stirred in 20 mL of water until complete dissolution. This solution was then added to the cellulose suspension, which was mechanically stirred and maintained at 20°C. The NaOCl (1.76 M solution, 11.5 mL, 20.24 mmol) was added dropwise to maintain the pH at 10 during the addition. After the total addition of NaOCl, the pH was maintained constant at 10 by adding 0.5 M NaOH solution until no more variation was observed indicating the end of the reaction. Methanol (5 mL) was then added to destroy the residual NaOCl and the pH was adjusted to 7 with 0.5 M HCl. A white suspension was obtained resulting from the oxidation of cellulose. After centrifugation, the supernatant, corresponding to the water-soluble oxidized cellulose with glucopyranose units completely oxidized, was separated to the water-insoluble fraction which corresponds to partially oxidized cellulose. The supernatant, referred to as polyglucuronic acid, was precipitated by adding an excess of ethanol, centrifuged, re-dissolved in water, dialyzed against water, and finally freeze-dried. The final yield of this process was between 90 and 95%. The water-insoluble fraction was further purified by successive centrifugation, re-dispersion in water and finally by dialysis against distilled water to obtain the nanocrystal suspensions. Its final yield was about 92-95% for cotton linters. In the text, the samples are named surface carboxylated cellulose nanocrystals.

GRAFTING REACTION

The grafting reaction between oxidized cellulose substrates and amines was performed according to Bulpitt and Aeschlimann [18] with minor modifications [22]. The typical run was achieved in aqueous media under stirring, with a typical addition of 2.5 mmol amine / 1 mmol glycosyl unit (2,5 eq). The pH of the suspension was adjusted to 7.5-8 with 0.5 M HCl. N-(3-Dimethylaminopropyl)-N'-ethylcarbodiimide hydrochloride (EDAC) and N-hydroxysuccinimide (NHS) with ratios of 1.5 relative to glycosyl unit [22] diluted in 2 mL of water were added (1,5 eq). The pH was adjusted and maintained to 8 by adding 0.5 M HCl and 0.5 M NaOH solutions. The suspension was stirred during 24 h at 50°C and finally precipitated by adding an excess of ethanol. After filtration on a 0.5 μm membrane and washing with ethanol (3 times), the precipitate was re-dispersed in water, evaporated again to remove any trace of ethanol, re-suspended in water and finally freeze-dried.

CHARACTERIZATION

Conductometric Titration

The residual carboxyl groups content of oxidized cellulose and grafted oxidized cellulose samples named degree of oxidation Do was determined by conductometric titrations [13]. Each dried cellulose (30-50 mg) was dissolved into 15 mL of 0.01 M hydrochloric acid solution. After 10 min of stirring (time to get stable suspension), the suspension was titrated with 0.01 M NaOH.

The titration curves, exemplified in Figure 1, showed the presence of a strong acid corresponding to the excess of HCl and a weak acid corresponding to the carboxylate content.

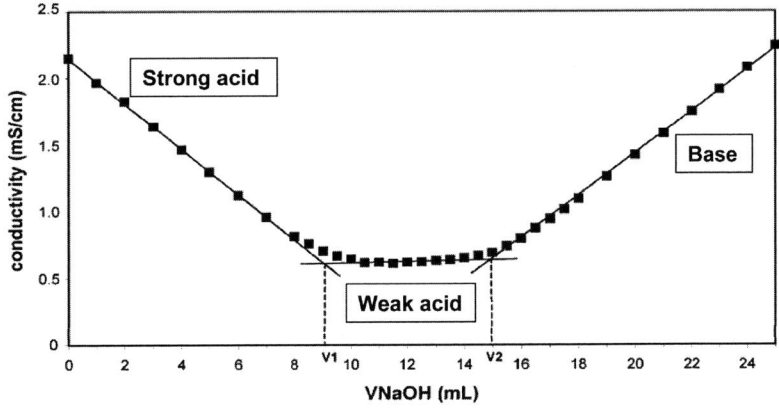

Figure 1. Typical conductometric titration curve of oxidized cellulose nanocrystals sample.

The carboxyl groups content of oxidized cellulose samples or degree of oxidation Do is determined by:

$$Do = \frac{162 \times n}{m - 36 \times n} \qquad (1)$$

After the grafting reaction, the residual carboxyl groups content Do_1 is determined by:

$$Do_1 = \frac{(162 + (Ma - 4) \times Do) \times n}{m + (Ma - 40) \times n} \qquad (2)$$

with 162 (g/mol) corresponds to the molar mass of an unreacted cellulose unit
Ma the molar mass of the amines
m the weight of oven-dried sample (g)
n (mol) the carboxylate content resulting to : $n = (V_2 - V_1) \times c$ (3)

where V_1 and V_2 are the equivalent volumes of NaOH (in L) and c is the exact concentration of NaOH solution (mol/L).

The degree of conversion (DC) which represents the number of grafted anhydroglucose units is deduced by the following equation: $DC = Do - Do_1$.

Infrared Spectroscopy

Infrared spectra were recorded on a FT-IR Perkin-Elmer 1720X spectrometer. Samples were studied as KBr pellets (1% in anhydrous KBr). Spectra were recorded using 3600 cm^{-1} spectral width (between 400 and 4000 cm^{-1}), 2 cm^{-1} resolution, and 32 scans were accumulated. Samples were studied as acidic form to avoid the superposition of sodium carboxylate peak with hydrogen bonds. For this, few milligrams of sample were suspended in 1 mL of water, 1-2 drops of 1M HCl were added and after stirring during 3-5 min the suspension was centrifuged and the precipitate was washed several times with water until neutrality.

Liquid-State NMR Spectroscopy

^{13}C and ^{1}H spectra were recorded with a BRUKER Avance 400 spectrometer operating at a frequency of 100.618 MHz for the ^{13}C and 400.13 MHz for the ^{1}H. Samples were studied as their sodium salt solutions in D$_2$O

(6-10 mg in 500 µL of solvent) at 30°C in 5 mm o.d. tubes. ^{13}C spectra were recorded using 90° pulses, 20000 Hz spectral width, 65536 data points, 1.638 s acquisition time, 2 s relaxation delay. From 4096 up to 10240 scans were accumulated depending on the sample solubility. Proton spectra were recorded with 4006 Hz spectral width, 32768 data points, from 4.089 up to 7.497 s acquisition time, 0.1 s relaxation delay and up to 128 scans. The 2D ^{13}C-^{1}H experiments were performed with 4006 Hz spectral width, 2048 data points, 0.266 s acquisition time, 1 s relaxation delay and 128 scans.

Solid-State NMR Spectroscopy

The NMR experiments were performed with a Bruker Avance 400 WB spectrometer operating at a ^{13}C frequency of 100.62 MHz using the combined technique of proton dipolar decoupling (DD), magic angle spinning (MAS) and cross-polarization (CP). ^{13}C and ^{1}H field strengths of 100 kHz were used for the matched spin-lock cross-polarization transfer. The spinning speed was set at 12,000 Hz for all the samples. The contact time was 2 ms, the acquisition time 30 ms and the recycle delay 4 s. The deconvolution of the spectra was achieved following earlier procedure [35]. The position and width of the lines were maintained constant throughout a series of samples. The area corresponding to the integration of the C1 signal was set to one. The evaluation of the oxidation and degree of crystallinity was made from the integration of the corresponding deconvoluted lines.

Electron Paramagnetic Resonance (EPR)

EPR measurements were made with a Bruker EMX X-band continuous wave spectrometer equipped with a Bruker ER 4116 DM rectangular cavity operating at 9.658 GHz. Experiments were performed at room temperature (300K) with a hyper frequency power of 1 mW and a modulation amplitude of 0.5 G. The amplitude of the magnetic field modulation and microwave power were adjusted so that no line-shape distortion was observable. The received gain was 63,200 and the sweep time was 42 s. Absolute quantification was obtained by comparison with a TEMPO sample of known concentration after double integration of EPR spectra.

Samples (2mg), i.e. 4-amino TEMPO, polyglucuronic acid-4-amino TEMPO and surface carboxylated cellulose nanocrystals-4-amino TEMPO, were dissolved in H$_2$O (1mL) and loaded into a closed capillary tube (o.d. 0.7 mm) which was introduced in a standard EPR tube (o.d. 3 mm).

In order to determine the degree of conversion DC from EPR data, taking into account cellulose substrates, the average molecular weight M of a glycosyl unit from derivatives was calculated. The number of 4-amino TEMPO ($n_{\text{4-amino TEMPO}}$) grafted with cellulose samples, was determined.

– with polyglucuronic acid sample:

$$M = 329 \times DC + (1 - DC) \times 198 = 131 \times DC + 198 \tag{4}$$

$$n_{\text{4-aminoTEMPO}} = \frac{DC \times m}{M} = \frac{DC \times m}{131 \times DC + 198} \tag{5}$$

which gives

$$DC = \frac{-198 \times n_{\text{4-aminoTEMPO}}}{131 \times n_{\text{4-aminoTEMPO}} - m} \tag{6}$$

– with surface carboxylated nanocrystals sample:

$$M = 198 \times Do_1 + (162 \times (1 - Do)) + (158 + Ma)(Do - Do_1)$$
$$M = 36 \times Do + 131 \times DC + 162 \tag{7}$$

$$n_{\text{4-amino TEMPO}} = \frac{DC \times m}{M} = \frac{DC \times m}{36 \times Do + 131 \times DC + 162} \tag{8}$$

which gives

$$DC = \frac{(-36 \times Do - 162) \times n_{\text{4-amino TEMPO}}}{131 \times n_{\text{4-amino TEMPO}} - m} \tag{9}$$

with 329 (g/mol) corresponds to the molar mass of a glucuronic unit grafted to 4-amino TEMPO

162 (g/mol) to the molar mass of an unreacted cellulose unit

198 (g/mol) to the molar mass of oxidized cellulose sodium salt (-COONa)

(158+Ma) to the molar mass of glycosyl unit grafted to 4-amino TEMPO (Ma = 171.26 g/mol)

m to the fraction of the dissolved sample which was introduced into capillary tube

X-Ray Diffraction

X-ray measurements were made on dried pellets of cellulose products. The X-ray diagrams were recorded on a Warhus vacuum flat plate X-Ray camera mounted on a Philips PW 1720 X-ray generator operated with Cu Kα radiation at 20 mA and 30 kV. Diffraction images were converted into 2θ-intensity profiles using specific software.

Elemental Analysis

The nitrogen content of the grafted samples was determined by elemental analysis. Indeed, the degree of conversion (DC) can also be calculated from the nitrogen content by using the following equation:

$$DC = \frac{36 \times Do + 162}{\frac{14 \times 100}{\%N} - M_a + 40} \tag{10}$$

where %N corresponds to the nitrogen content of the grafted samples

162 (g/mol) to the molar mass of cellulose unit

Ma the molar mass of 4-amino TEMPO

Chapter 3

RESULTS AND DISCUSSION

Before to dig into the core of this work, it is worth considering the cellulose as the main substrate, recalling the knowledge about its structure and underlining some of its organization principles which are key points for all chemical modifications.

1.) ORGANIZATION AND CHEMICAL STRUCTURE OF CELLULOSE

Since cellulose is reproduced by nature with fixation of CO_2 gas, cellulose is the most abundant organic polymer on the Earth. Cellulose is the structural component of the primary *cell* of *green plants*, many forms of *algae* and the *oomycetes*. Some species of *bacteria* secrete it to form *biofilms* and also certain sea animals like tunicates. Cellulose is the most common organic compound on Earth. For an example, about 33 percent of all plant matter is cellulose.

Cellulose chains are organized into microfibrils which constituted the backbone of cell wall and presented a very high resistance to tension. The cellulose fiber is composed by macrofibrils themselves composed of bundles of microfibrils and each microfibril, in turn, is composed of bundles of cellulose chains (Figure 2). The cellulose *microfibrils* are linked via hemicellulosic tethers to form the cellulose-hemicellulose network (macrofibrils), which is embedded in the pectin matrix containing lignin and proteins. Primary cell walls characteristically extend (grow) by a mechanism called acid, which involves turgor-driven movement of the strong cellulose microfibrils within the weaker hemicellulose/pectin matrix, catalyzed by expansion proteins. The outer part of the primary cell wall of the plant epidermis is usually impregnated with cutin and wax, forming a permeability barrier known as the plant cuticle.

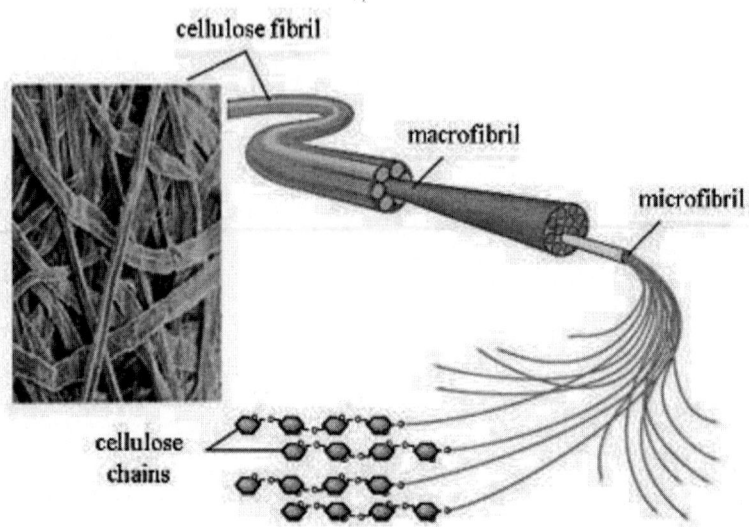

Figure 2. Structure of cellulose fiber.

The cellulose microfibrils represent about 20 to 30% of the dry mass of parietal material and occupy about 15% by volume of the wall. For the differentiate cells with a secondary wall, the proportion of cellulose reaches 40 to 90% of parietal mass.

The main sources of cellulose are issued to i) primitive organisms like bacteria (ex: Acetobacter xylinum), algae (ex: Valonia, Cladophora, Microdictyon); ii) plants (ex: wood, cotton, flax, ramie, jute, parenchyma of sugar beet pulp...); and iii) **envelop of sea animals** belonging to the Ascidians family (ex: tunicate).

1.1.) Chemical Structure

Cellulose was discovered in 1838 by the French chemist *Anselme Payen*, who isolated it from plant matter [36]. **From to Payen' works, the basic** cellulose formula has been determined by Weillstatter and Zechmeister [37-39]. Cellulose, *organic* with $(C_6H_{10}O_5)_n$ *formula*, is a linear homopolymer, consisting of several hundred to over ten thousand ß(1→4)-glycosidic linked D-glucopyranose units (Figure 3) with three hydroxyl groups on C2, C3 and C6 carbons [40]. The repetitive unit is named cellobiose which is constituted by two D-glucopyranose units placed at 180° giving the molecule its linear characteristic.

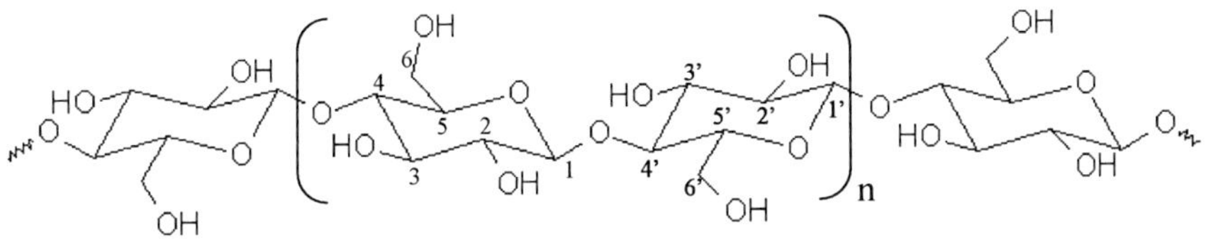

Figure 3. Schematic representation of a cellulose chain.

Crystallographic investigations of D-glucose and cellobiose [41] established unambiguously that the D-glucose residues had the $^4C_1$ chair conformation.

The two chain ends are chemically different: one end has a D-glucopyranose unit in which the anomeric carbon atom is involved in a glycosidic linkage whereas the other end has a D-glucopyranose unit with free anomeric carbon atom (Figure 4).

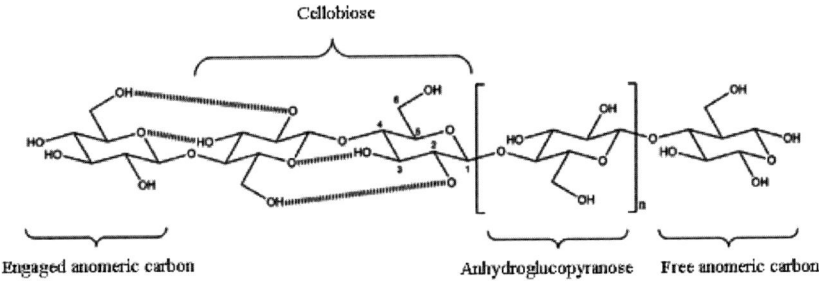

Figure 4. Chain ends of cellulose chain.

This cyclic hemiacetal function is in an equilibrium in which a small proportion is an aldehyde which gives rise to reducing properties at this end of the chain: the cellulose chain has a chemical polarity. Determination of the relative orientation of cellulose chains in the three-dimensional structure has been and remains one of the major problems in the study of cellulose. So, in the cellulose crystal, two arrangements are possible: either an organisation in parallel chains with reducing chain end placed in the same side or an organisation in antiparallel chains with alternate position.

1.2.) Crystallinity and Polymorphism of Cellulose

The $^4C_1$ chair conformation of D-glucopyranose units added to the ß(1→4)-glycosidic link induces to a straight chain polymer with no coiling or branching occurs, and the molecule adopts an extended and rather stiff rod-like conformation, aided by the equatorial conformation of the glucose residues.

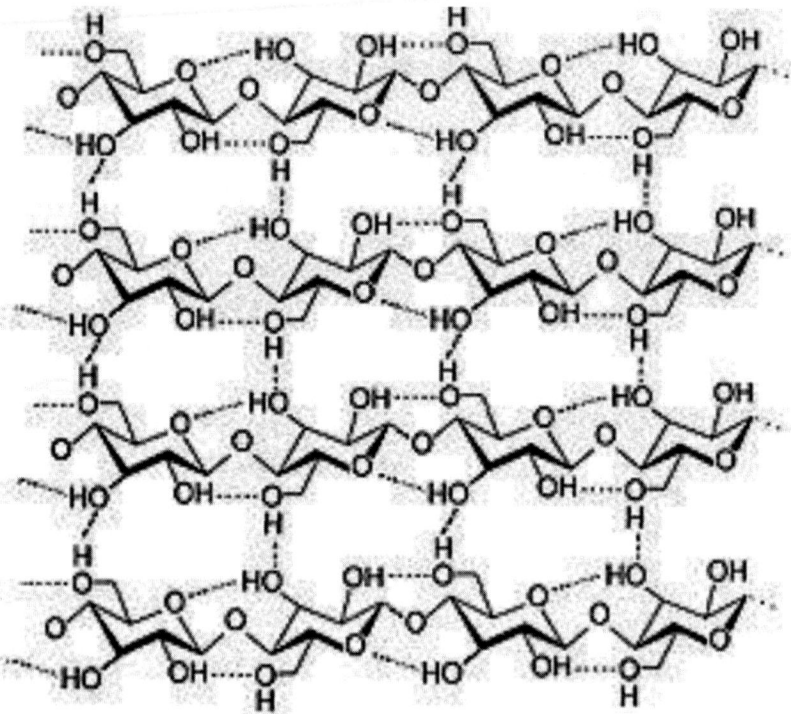

Figure 5. Schematic representation of cellulose chains showing the hydrogen bonds (dashed).

The free hydroxyl groups present in the cellulose are likely to be involved in a number of *intra* and *inter* molecular hydrogen bonds which may give rise to various ordered crystalline arrangements (Figure 5). The *intra* molecular hydrogen bonds are realized between i) hydrogen O3-H and heterocyclic oxygen O5 on a neighbor cycle and ii) hydrogen O2-H and oxygen O6-H on a neighbor cycle (Figure 6). The *inter* molecular hydrogen bonds are performed with oxygen O3-H and primary hydrogen O6-H on a neighbor chain (Figure 6).

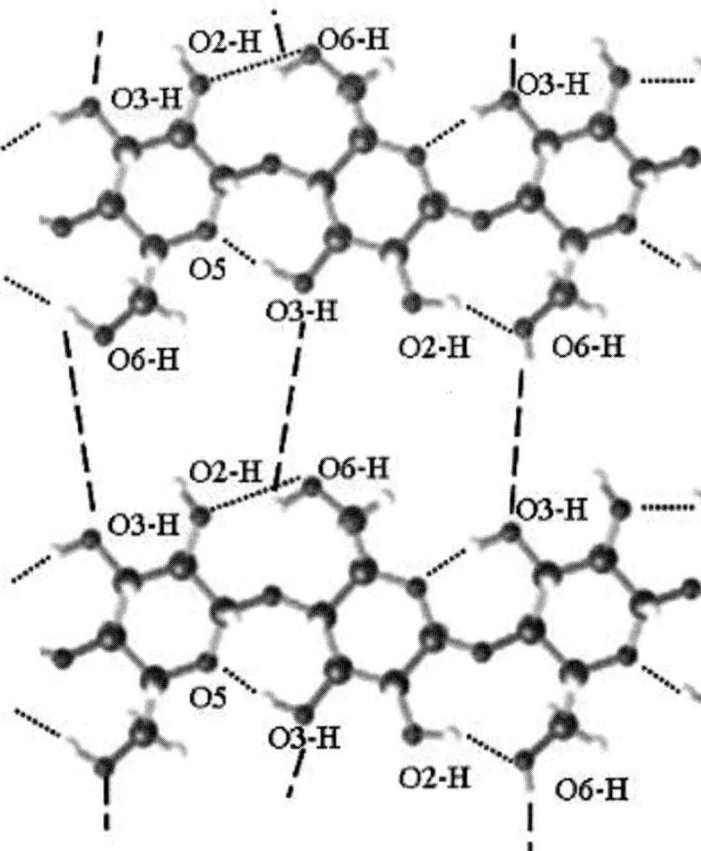

Figure 6. Representation of *intra* and *inter* molecular hydrogen bonds into cellulose.

The *intra* molecular hydrogen bonds confer the rigidity to cellulose chain whereas the *inter* molecular hydrogen bonds ensure the cohesion of crystalline structure formed by the all cellulose chains. The combination promotes the alignment in parallel beams leading to regular arrangement of rigid molecules.

In the case of cellulose, these crystalline arrangements are usually imperfect, in terms of crystallinity, crystal dimensions, chain orientation and then the purity of the crystalline form must be taken into consideration. The crystal density can be gauged from the crystallographic data which leads to suggest the importance of the amorphous components generally present. By infrared spectroscopy, the degree of crystallinity can also be estimated as a function of the relative intensity of specific bands [42].

Four principal allomorphs have been identified for cellulose by its characteristic X-ray diffraction pattern: I, II, III and IV [43]. The relationships among the various allomorphs are shown schematically (Figure 7). The natural form of cellulose, called cellulose I or native cellulose, is the most abundant form. Its highly complex and not yet completely resolved three-dimensional structure results of co-existence of two distinct crystalline forms cellulose I_α and I_β. This was a major discovery and led to a revival of interest in the study of cellulose structure [44-45]. Cellulose II, named regenerated cellulose, is generally obtained by regeneration of cellulose from solution or by mercerization (alkaline medium) [46]. The transition from cellulose I to cellulose II is not reversible and this implies that cellulose II is a stable form compared with the metastable cellulose I.

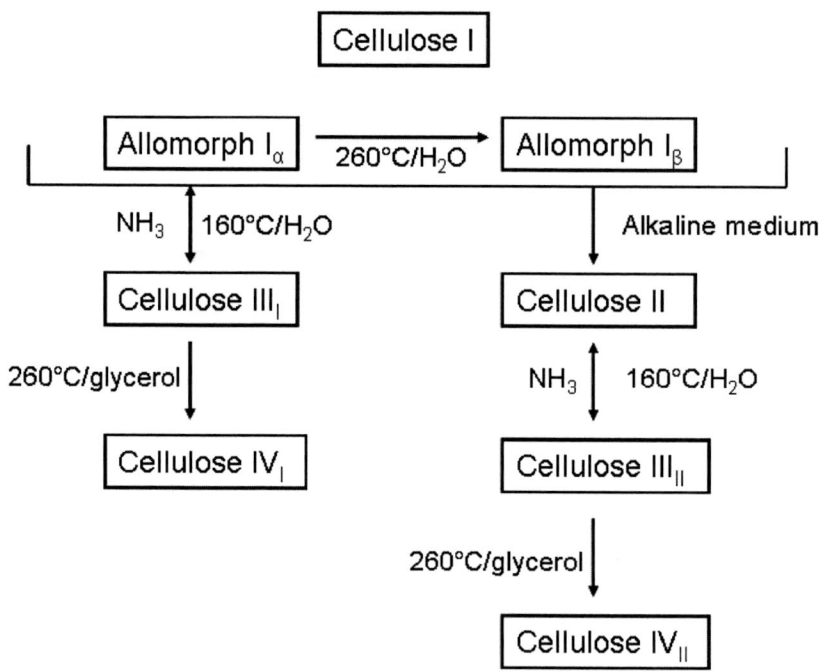

Figure 7. Relationship between various allomorphs of cellulose.

The preparation of cellulose III is realized with ammonia treatment and allows to obtain the form cellulose III_I or the form III_{II} from to cellulose I and cellulose II, respectively. Cellulose III can be transformed into cellulose IV after treatment at high temperature in glycerol: cellulose IV_I and IV_{II} obtained from cellulose III_I and III_{II}, respectively. After relative recent studies by electron diffraction [47], it is now accepted that cellulose IV_I is a disordered form of cellulose I (native cellulose form). This allomorph, at native state, may be observed in some mushrooms but also in primary cell walls of cotton [48].

1.3.) Crystalline Structures

The X-rays diffraction is the oldest technique used to characterize the crystallinity of cellulose. First work suggested that native celluloses with different origins seem to crystallize in different arrangements whereas from *Valonia* and bacterial sources, cellulose presents a same crystalline unit cell. The crystalline structure of cellulose is refined with new techniques based on solid-state ^{13}C NMR spectroscopy which confirmed the existence of two families of native cellulose [44-45]. From a detailed analysis of the carbon atom couplings, Vanderhaart and Atalla [44-45] established that native cellulose corresponded to a composite of two distinct crystalline phases named a one-chain triclinic structure I_α and a two-chain monoclinic structure I_β. The fractions of I_α and I_β crystalline phases in any native cellulose samples depend on the origin of the cellulose. This model was supported by electron diffraction study of native cellulose from algal cell wall [49] and by computational prediction [50]. In general, the celluloses produced by primitive organisms (bacteria, algae etc.) are enriched in the I_α phase whereas the cellulose of higher plants (woody tissues, cotton, ramie...) consists mainly of the I_β phase. Indeed, the crystallographic studies of envelop of sea animals like tunicate show crystalline form uniquely composed of the I_β phase [51].

The discovery of the crystalline dimorphism of cellulose was the starting point for a number of research projects of which the aim was to evaluate the properties of each allomorph and procedures for their interconversion. Details of the crystalline structure of these two forms were reported by Kono [52] using ^{13}C NMR technique and by Nishiyama [53-54] using synchrotron X-ray and neutron fiber diffraction. The estimation of phase's composition of native cellulose is possible using different techniques such as FTIR [55], ^{13}C NMR [56-57] and synchrotron-radiated X-ray diffraction [58]. The I_α phase is a metastable from which can be converted to the more stable I_β form by annealing in different medium [59-60]. The Table 3 and Figure 8 report crystallographic characteristics of allomorphs of native cellulose. The crystalline forms of cellulose II, III and IV are organized in monoclinic unit cell as observed for cellulose I_β whereas the cellulose I_α is oriented in triclinic unit cell [53].

Table 3. Crystallographic characteristics of allomorphs of cellulose.

Allomorph	Unit cell	a (Å)	b (Å)	c (Å)	Angles (°)
I_α	Triclinic	6.74	5.93	10.36	$\alpha = 117$ $\beta = 113$ $\gamma = 97.3$
I_β	Monoclinic	8.01	8.17	10.36	$\gamma = 97.3$

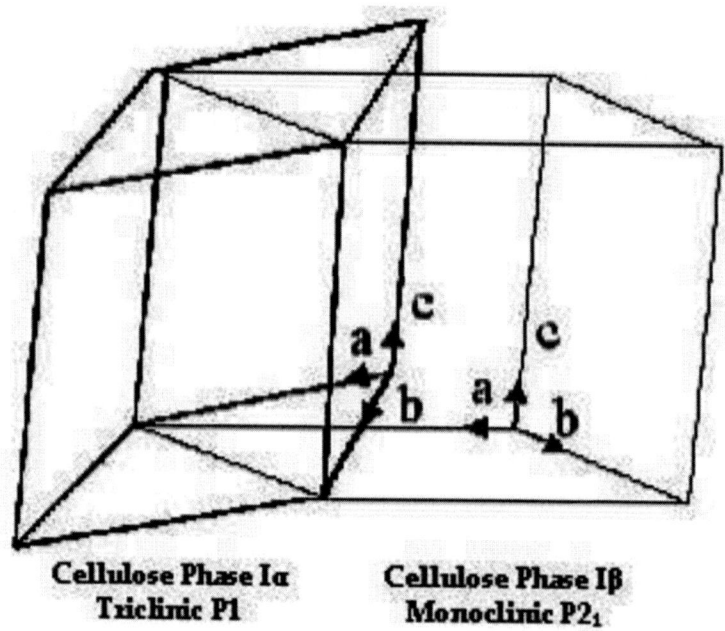

Cellulose Phase I_α
Triclinic P1

Cellulose Phase I_β
Monoclinic P2₁

Figure 8. Orientation of unit cells of monoclinic cellulose I_β and triclinic cellulose I_α.

Each step of interconversion induces change in crystallographic features of cellulose and in chemical assignment of C1, C4 and C6 specific carbons (Table 4 and Table 5).

Table 4. Chemical shifts of C1, C4 and C6 carbons characteristic of cellulose allomorphs.

Allomorph	Chemical shifts (ppm)		
	C1	C4	C6
Cellulose I_α	106.0-106.3	89.8-90.2	66.2-66.5
Cellulose I_β	105.0-105.2	89.1-89.3	65.5-66.1
Cellulose II	105.8-106.3	88.7-88.8	63.5-64.1
Cellulose III_I	105.3-105.6	88.1-88.3	62.5-62.7
Cellulose III_{II}	106.7-106.8	88.0	62.1-62.8
Cellulose IV_I	105.6	83.6-84.6	63.3-63.8
Cellulose IV_{II}	105.5	83.5-84.6	63.7
Amorphous	105	108	63

Table 5. Characteristic diffraction angles of cellulose allomorphs.

Allomorph	2θ Diffraction angle (°)			
	110	110	200	012
Cellulose I	14.8	16.3	22.6	-
Cellulose II	12.1	19.8	22.0	-
Cellulose III_I	11.7	20.7	20.7	-
Cellulose III_{II}	12.1	20.6	20.6	-
Cellulose IV_I	15.6	15.6	22.2	-
Cellulose IV_{II}	15.6	15.6	22.2	20.2

The mercerization of native cellulose, consisting on swelling of cellulose in alkaline medium following by washing with water, induces that the conversion from cellulose I to cellulose II is as a function of crystallinity of cellulose, NaOH concentration, time and temperature of reaction. The cellulose II presents an antiparallel arrangement of cellulose chains inside the unit cell of crystal. This arrangement allows the establishment of more number of hydrogen bonds than in native cellulose explaining also a better stability of this allomorph.

Concerning the cellulose III, numerous studies were focus on reversible conversion from cellulose I to cellulose III based on the use of electronic microscopy [61-62], solid-state NMR [29] or X-ray diffraction [63]. It is interesting to note that the conversion from cellulose I to cellulose III_I, in the case of Valonia cellulose, was accompanied by an important de-crystallisation and fragmentation of the cellulose crystal. The reverse transition resulted in partial re-crystallization but this did not allow complete restoration of the damage done to the morphological surface. Characterization by electron diffraction revealed that the uniplanar-uniaxial orientation of the crystalline cellulose microfibrils was lost completely during the stage of swelling and washing necessary for the conversion into cellulose III_I. Washing with methanol resulted in the formation of irregularities into which were inserted crystalline domains of small dimensions. The final material which crystallized in the cellulose I form was obtained by treatment with hot water and characteristically displayed an increase in the accessible surface and consequently reactivity.

In spite of some minor differences the results agree sufficiently well to propose a model in which the cellulose chains has almost perfect two-fold symmetry and is compatible with occurrence of two intermolecular hydrogen bonds between consecutive residues.

1.4.) Microfibrils of Native Cellulose

The microfibrillar structure of cellulose has been established beyond doubt through the application of electron microscopy [64-65] and great variations in dimensions, depending on origin, have been reported [47-48, 66]. The application of transmission electron microscopy [67-72] has established with certainty that the microfibril is the basic crystalline element of native cellulose [1, 67, 68-70, 73]. Different levels of structural organisation of cellulose are now well characterized.

An experimental protocol using *exo*-cellulases finally reached the proof of parallel arrangement in the family of native cellulose [74]. Recent investigations using complementary enzymatic and chemical staining of reducing ends have supported this model [75] and, at the same time, produced precise descriptions of the orientation of the chains relative to the crystal axes. Hence the crystalline microfibrils possess the same polarity as the chains of which they are composed.

In addition to the crystalline phases, native cellulose contains disordered domains which can be considered like amorphous. Native cellulose may be assigned to a semi-crystalline fibrillar material. Presence of disordered phases was supported by experimental results from solid state ^{13}C NMR characterization, tensile testes of cellulose fibers, wide-angle (WAXS) and small-angle X-ray scattering (SAXS).

1.5.) Whiskers of Cellulose

Depending on their origin, the microfibril diameters range from about 2 nm to 20 nm for lengths that can reach several tens of microns (Figure 9).

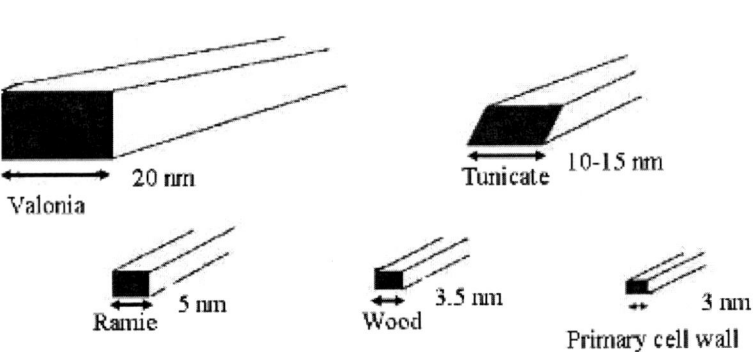

Figure 9. Schematic representation of range of microfibril size from different sources.

As they are devoid of chain folding and contain only a small number of defects, each microfibril can be considered as a string of cellulose crystals, linked along the microfibril by amorphous domains (Figure 10).

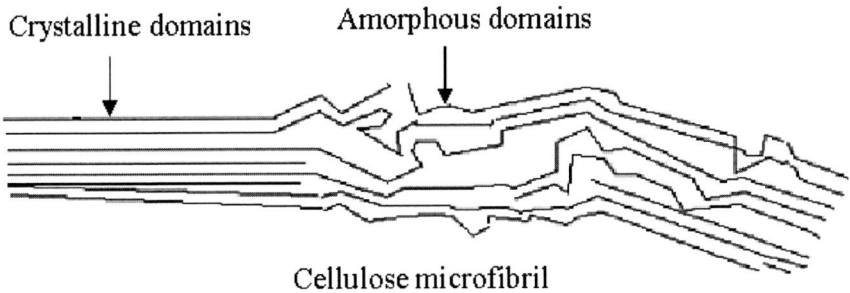

Figure 10. Schematic representation of amorphous and crystalline domains of cellulose microfibril.

Some imperfections arose from dislocations at the interface of microcrystalline domains along the microfibril length. These imperfections were used to advantage by treatment with acid to produce monocrystals called "whiskers" having the same diameter as the starting microfibrils but much shorter length. These monocrystals corresponded to nearly perfect crystalline arrangement. After hydrolysis, the monocrystals of cotton linters are shorter than those issued to long microfibrils like Valonia cellulose. These microcrystals reach lengths of 0.1 micron and widths from 10-50 Å and have a shape value of 20. At the other end of the spectrum, cellulose microfibrils from parenchyma (low crystalline cellulose) are produced by a mechanical treatment which, contrary to the hydrolysis, allows disruption of the microfibrils without affecting the original length. As a result, microfibrils several microns long and 20-30 Å wide are obtained. It follows that microfibrils with high section (~ 25 nm) presented few per cent of amorphous domains whereas for cotton with 5 nm section and for primary cell wall like sugar beet pulp with 3 nm section, the amorphous domains corresponded to about 30% and 70%, respectively [1]. The smaller microfibrils, the more surface chains are and the more non-crystalline component is rising correspondingly.

1.6.) Cellulose as Starting Materials for Grafting Purposes

Currently, many search laboratories are involved in cellulose research and development for either industrial use or structural knowledge increase. To precise crystallographic description remains a fundamental target.

Generally, for grafting reaction, it is used cotton cellulose that is a fiber issued to cotton plant for which the stem is covered with flowers to 5 lobes. The benefits arising are ovoid capsules that open to release seed wrapped a white duvet called Linter.

In order to obtain cellulose **monocrystals ("whiskers"), an acid** treatment of cellulose microfibrils is advocated. The amorphous regions act as structural defects and are responsible of the transverse cleavage of the microfibrils into short monocrystals under acid hydrolysis. This procedure largely described is used to prepare highly crystalline cellulose particles. Under controlled conditions, this transformation consists of the disruption of amorphous regions surrounding and embedded within cellulose microfibrils while leaving the microcrystalline segments intact. Highly crystalline cellulose consists generally of a stiff rod-like **particle called "whiskers" (or hydrolyzed** nanocrystals). Geometrical characteristic of cellulose nanocrystals depend on the origin of cellulose microfibrils and acid hydrolysis process conditions such as time, temperature and purity of materials. The most studied cellulose sources are: Valonia [76], cotton linters [77], wood pulp [78] and sugar beet pulp [35, 79].

Two acid treatments can be possible to form nanocrystals [80]. The use of sulfuric acid hydrolysis leads to stable aqueous colloidal suspension which could form chiral nematic ordered phases when the suspension concentrations exceed a critical value, and the nanocrystals present a negatively charged surface (sulfate esters groups). The cellulose nanocrystals suspensions resulting from hydrochloric acid hydrolysis have minimal surface charges and are organized into small aggregates.

2.) OXIDATION REACTION OF CELLULOSE SAMPLES

2.1.) Choice of Oxidation Reaction

Oxidation of polysaccharides has been studied in detail by numerous investigators, but, because of the presence of several reactive groups, it is not easy to attain high selectivity, and only a limited number of reagents are available for this purpose [81]. Because it is insoluble in water and most common organic solvents, there are especially difficulties with cellulose. In view of the structure of an anhydroglucose unit of cellulose, one has to account for the presence of three reactive groups available for oxidation: one primary (C6) and two secondary (C2, C3) hydroxyl groups.

Direct oxidation of hydroxyl groups provides interesting routes for the introduction of carbonyl and carboxyl groups into polysaccharides. Oxidation of secondary hydroxyl groups has been achieved by various oxidation reagents, e.g., periodate and hypochlorite, which results in oxidative scission of 1,2-diols and the formation of dialdehyde and dicarboxylic structures, respectively [82-83] (Figure 11).

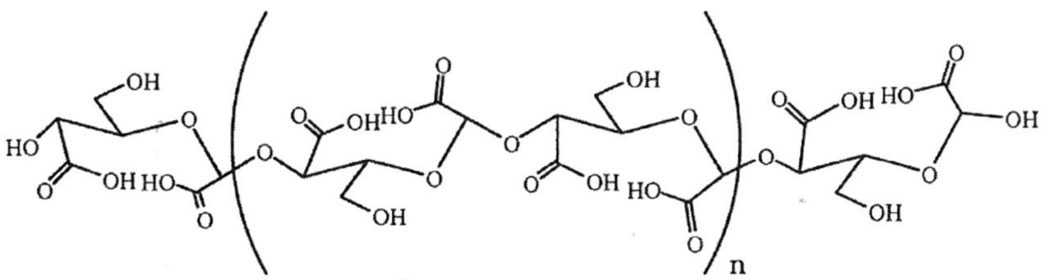

Figure 11. 2, 3-dicarboxy derivative after sodium chlorite treatment.

Full oxidation of monosaccharides by nitric acid to aldaric acids has been an established technique for more than a century. Analogously, oxidation of polysaccharides such as cellulose or starch by nitrogen dioxide (N_2O_4) yields 6-carboxy starch and 6-carboxycellulose respectively [84]. Subsequent hydrolysis at rigorous conditions (0.5-2M HCl at 150°C) of these materials yields D-glucuronic acid. A drawback of the oxidation with nitrogen dioxide is that depolymerisation may be an important side reaction. An improvement of the process with respect to this aspect can be achieved by conducting the reaction when the polysaccharide is dissolved in 85% phosphoric acid and with sodium nitrite as the oxidant: better yield versus a more important depolymerisation [85-86].

In the recent decade, catalytic oxidation of carbohydrates transposed to polysaccharides using highly regio-selective and stable nitroxyde radical (2,2,6,6-tetramethylpiperidine-1-oxyl radical (TEMPO)) has become one of the most promising procedures to convert polysaccharides into the corresponding polyuronic acids. The method is very suitable for selective oxidation of primary hydroxyl groups into aldehydes and / or carboxylic acid groups. Contrary to enzymatic or metal-catalyzed oxidation, the TEMPO mediated oxidation is highly effective in the conversion of high molecular weight polysaccharides. Other advantages to be mentioned in connection to the TEMPO-oxidation process are the high reaction rate and yield, the high selectivity, the catalytic process, just a modest degradation of polysaccharides throughout the process. This process was first utilized some 15 years ago, but to this day, to our knowledge, no commercial process is in operation. After the first publications on TEMPO mediated oxidation of polysaccharides and other types of alcohols, numerous papers have been published.

2.2.) Mechanism of Reaction

The oxidation of primary alcohol groups in natural polysaccharides, catalyzed by 2,2,6,6-tetramethylpiperidine-1-oxyl radical (TEMPO) has been recently proposed as a more selective, faster and better-controlled method [3, 5-6] as opposed to the traditional procedure using nitrite/nitrate in concentrated phosphoric acid [84, 87]. TEMPO and its analogues belong to a class of compounds which usually are referred to as nitroxide radicals. This compound is secondary amine nitrogen oxide with the general chemical structure given below (Figure 12).

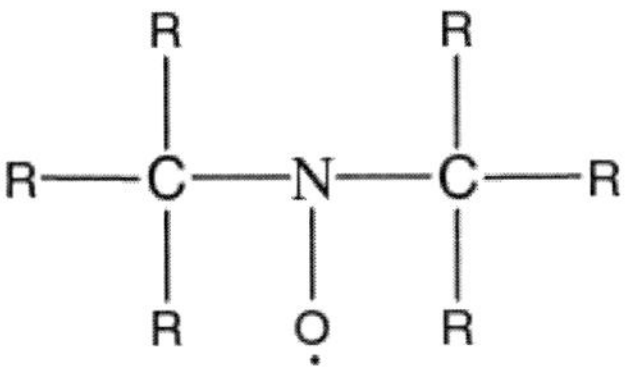

```
         R           R
         |           |
    R —— C —— N —— C —— R
         |    |    |
         R    O    R
              ·
```

Figure 12. General chemical structure of nitroxide radicals.

2,2,6,6-tetramethyl-1-piperidine oxoammonium radical (1) is known to be a stable radical because of the unpaired electron delocalized between the nitrogen and the oxygen atom. The NO-bond energy has been estimated to be 420 kJ/mol which approximately corresponds to the energy of one and a half bond. The most striking feature of nitroxyde radicals is their high or extreme stability toward dimerization or decomposition and inertness to typical organic molecules. However, the stability largely depends on the structure of the radical compound and especially on the nature of the substituents attached next to the nitroxyl group. For instance, hydrogens attached on the α and α'-carbons are a source of instability leading often to disproportionate the radical. About cyclic nitroxide radical such as TEMPO, the presence of electro-donor methyl groups on carbons indicated an exceptional stability and inertness. That induced high redox properties help characterize the radical TEMPO [4]. The cyclic nitroxide radical of the piperidine is interrelated by one-electron transfer oxidation or reduction reactions: either the radical can be oxidized by several reagents (oxidant) to give a nitrosonium ion (2) [88] or reduced to give the hydroxylamine (3) by reducing agents (Figure 13). In all studies, the nitrosonium ion is assumed to be the responsible oxidizing specie. The oxidation potential of TEMPO is not negligible, but in order to function as efficient oxidants of organic substrates, it converted into the more powerful nitrosonium ion. The nitrosonium ion shows selectivity towards primary hydroxyl functions over secondary ones.

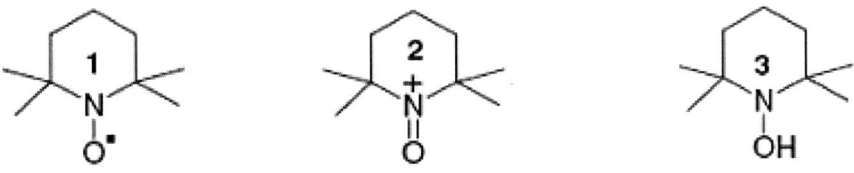

Figure 13. Nitroxyde radical TEMPO.

In most of works, TEMPO radical is utilized as oxidation catalysts: the nitrosonium ion is continuously regenerated *in situ* by a primary oxidant. The main advantage based on that the radical can be added in catalytic amounts. The co-oxidant must be inert towards organic substrates to oxidize. The procedure for the *in situ* generation of TEMPO was first described by Semmelhack [89].

Following the method developed by Davis and Flitsch [90] using a mixture of sodium hypochlorite, sodium bromide and 2,2,6,6-tetramethyl-1-piperidine oxoammonium radical, the TEMPO-NaBr-NaClO system was applied to a wealth of products including many polysaccharides. This method was first proposed for water-soluble polysaccharides [3, 5-8], namely starch, inulin, amylodextrin, pullulan, alternan, amylopectin, chitosan, galactomannan, and later extended to water-insoluble products [7-12], such as cellulose, amylose, and chitin.

The principle of the catalytic system involved TEMPO in presence of two primary co-oxidants: sodium hypochlorite and sodium hypobromite. The reactional mechanism is presented in Figure 14.

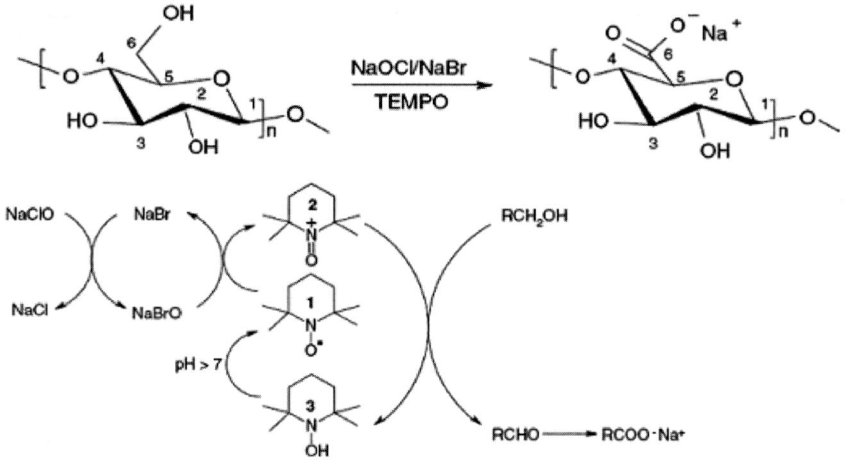

Figure 14. Reactional mechanism of TEMPO mediated oxidation.

The catalytic mechanism of TEMPO mediated oxidation is based on the regeneration in turn of nitrosonium ion (2) in presence of regenerating oxidant which is NaOBr (Figure 14). NaOBr is the product of oxidation of NaBr by NaOCl (primary co-oxidant) and then reacted on TEMPO radical to give the nitrosonium ion. At alkaline pH, the nitrosonium ion is continuously regenerated throughout the conversion of hydroxylamine (3), which is formed after substrate oxidation, into the radical (1). The nitrosonium ion reacted with primary hydroxyls of polysaccharides to get aldehyde groups releasing the hydroxylamine (3). In turn, the aldehyde was modified in carboxylic acid with the same principle. So, two equivalents are necessary during the reaction: the former to obtain aldehyde groups and the latter to converse the aldehyde groups to carboxylic acid groups.

The products obtained by this procedure present lower molecular weight than those obtained via a two step method with for an example $NaIO_4$ and $NaClO_2$ (methods known for the glycolic oxidation of polysaccharides) inducing the perspectives for biodegradability are better [91]. Another advantage is that the product will be cheaper since it is prepared in a single, simple step with a relatively inexpensive reagent.

For native cellulose (allomorph I of cellulose), the oxidation led to a two-phase solution: water-soluble and water-insoluble fractions. These two fractions corresponded to the totally oxidized polyglucuronic acid and the surface carboxylated cellulose nanocrystals, respectively. The oxidation runs were carried out according to the above-mentioned scheme. Sodium hypochlorite, used to regenerate the catalyst, is added dropwise during the reaction. It has been shown that a slow addition of the oxidant led to a higher yield and products that were less coloured (so less degraded!). The sodium bromide allowed to the regeneration of sodium hypochlorite. The oxidation must be realized at pH rigorously kept at 10. Indeed, with higher pH, a severe degradation by β–elimination of cellulose occurred. With an acid pH, the secondary hydroxyl groups could be oxidized [4].

2.3.) Influence of Oxidation Parameters

Oxidant. The reagent sodium hypochlorite should be added gradually to avoid undesired reactions such as glycol cleavage due to the presence of excess sodium hypobromite [92]. Indeed, the amount of primary oxidant determines the yield of the oxidation reaction.

Solvent. Herrmann [93] reported on a water-based system (acetic acid – water) that only gave the aldehyde groups. These results suggest that the conversion of aldehydes to carboxyl groups predominantly takes place in the water phase. Since the presence of water is essential for the latter conversion [4], it is plausible that only the hydrated form of aldehyde intermediate can be oxidized further (Figure 15).

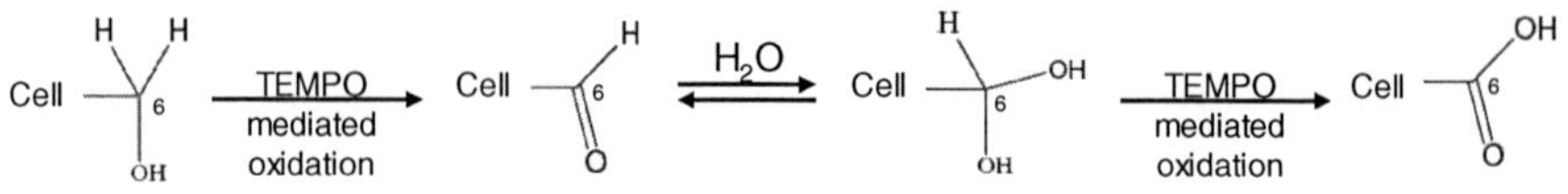

Figure 15. TEMPO mediated oxidation of primary alcohols *via* the hydrated aldehyde intermediate.

pH value. Significant for TEMPO mediated oxidation reactions of polysaccharides is the high chemoselectivity for the primary alcohol groups, which probably is due to the steric hindrance of the secondary hydroxyl groups. The researches on the oxidation of cellulose showed that both the rate of oxidation and the nature of the products are determined by the pH of the solution. It was established that in the TEMPO-NaBr-NaClO system, the pH optimum rate of TEMPO is found at pH 10 [13, 22, 95]. If the pH value is lower than 9.5, an undesired oxidation of secondary hydroxyls functions by hypobromite during the reaction is observed in competition with the oxidation of primary hydroxyls ones [5]. With higher pH than 10.5, a severe degradation by β–elimination of cellulose occurred [6].

Temperature: Low temperature during reaction is essential for minimizing the degradation of the polymer. In previous studies [94], the TEMPO-NaBr-NaClO system is usually carried out at 0-5°C (in an ice batch). To accelerate the reaction, the oxidation was to perform at increased but still moderate temperature (20°C) without loss of selectivity and presence of side reactions [22].

2.4.) Characteristics of Cellulose Substrates

In general, the evolution of the oxidation was followed by the sodium hydroxide consumed to neutralize the carboxylic acids generated in C6 carbons (conductometric titration). The increase of NaOCl concentration up to 2 eq involved an increase in the degree of oxidation (Do) of cellulose samples.

For partially oxidation leading to surface carboxylated cellulose nanocrystals, the TEMPO mediated oxidation led to approximately constant molar yields of oxidized nanocrystals more or less constant, 92–95% with cotton linters (for soluble fractions: 4–7%) (Figure 16).

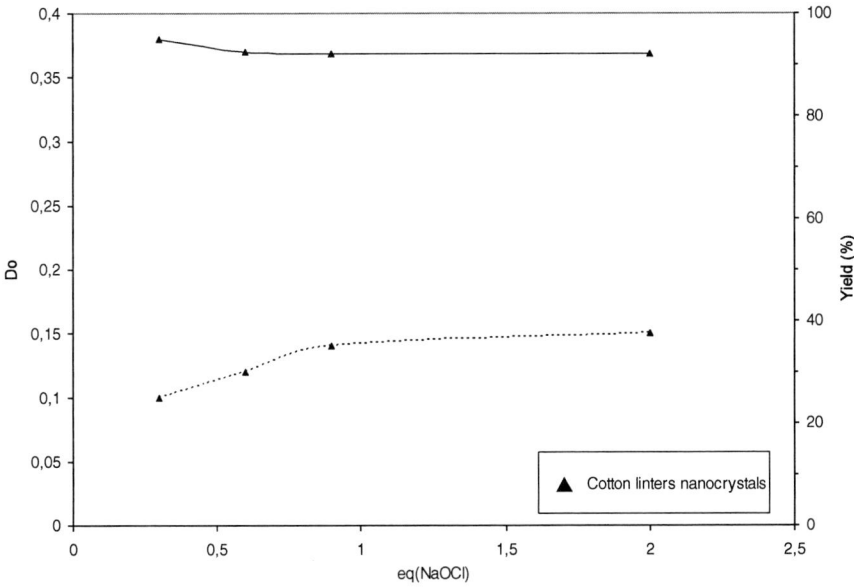

Figure 16. Evolution of degree Do (...) and yield (–) of oxidation versus the amount of added NaOCl for cellulose nanocrystal samples.

The global yields for hydrolyzed samples were always 100%. Montanari [95] indicated that the cellulose samples which are not hydrolyzed before oxidation presented lower global yields with increasing NaOCl concentrations. The values decreased to around 90% for cotton linters. The degradation of cellulose samples explained this difference between hydrolyzed and nonhydrolyzed samples. As a result, in this chapter, hydrolyzed cellulose samples are privileged: few loss of material correlated with no degradation and no side reactions.

A constancy in the Do value as of 1 equivalent of NaOCl is shown (Figure 16). This maximal value not corresponds to the theoretical value which would be 2 mmol of NaOCl/mmol of glycosyl unit (2 eq). This value depends on the crystal size and corresponds roughly to the amount of primary hydroxyl groups easily accessible to the reactants, *i.e.* surface hydroxyl groups. In addition, the Figure 16 showed efficient molar yields with the constancy in value indicating no degradation of cellulose samples with the NaOCl concentration.

From these observations, the TEMPO mediated oxidation of cellulose nanocrystals is performed with 1 mol NaOCl / 1 mol glycosyl unit (1 eq). The carboxyl content of surface carboxylated cellulose nanocrystals was determined by conductometric titrations with very reproducible results and the value is 0.15 for cotton linters (Figure 16).

For totally oxidized polyglucuronic acid (from cellulose III_I), a constant molar yield in the order of 100% is obtained. The conversion into cellulose III_I involved starting material which is rapidly and fully oxidized yielding polyglucuronan whereas for native cellulose, around 10% of water-soluble product is obtained. The accessibility improve is due to the increase in distance between chains in the crystal (confirmed by the density variation from 1.61 (cellulose I) to 1.54 (cellulose III_I)) and also the fact that hydroxyl groups are less involved in hydrogen bonds network. Similar results are shown in the literature [13, 28].

Table 6. The ^{13}C NMR characteristics of polyglucuronic acid (liquid-state NMR) and surface carboxylated cellulose nanocrystals (solid-state NMR).

Liquid-state ^{13}C NMR, δ (ppm)		Solid-state ^{13}C NMR, δ (ppm)	
C1	103.5	C1	105
C2	73.75	C2-3-5	70-77
C3	75.25	C4 amorphous	83.5
C4	81.95	C4 crystal	88.7
C5	76.30	C6 amorphous	62.4
C6 oxidized	175	C6 crystal	65
		C6 oxidized	to174.8

Table 7. FTIR characteristics of oxidized cellulose.

FTIR (in acid form) cm^{-1}			
OH	3340	Water adsorbed onto the surface of nanocrystals	1640
CH	2900		
C=O (free COOH)	1730		

The solid-state NMR characteristics of surface carboxylated cellulose nanocrystals and liquid-state chemical shift data of polyglucuronic acid are given in Table 6 with the main appearance of the peak at around 175 ppm and the FTIR peak at 1730 cm^{-1} corresponding to carboxyl moiety. The FTIR characterization of these oxidized cellulose samples is shown in Table 7.

The oxidation of hydroxyl moieties allows the presence of carboxylic functions, which can be grafted easily in order to get prodrug systems.

3.) GRAFTING REACTION

3.1.) Mechanism of Grafting Reaction

The aim of this part is to guide the chemical modification to the grafting reaction of amines into cellulose samples.

Commonly, grafting techniques of carboxylic derivatives largely used carbodiimides as grafting reagents with in particular N-(3-Dimethylaminopropyl)-N'-ethylcarbodiimide hydrochloride (EDAC). EDAC is a water-soluble condensing reagent generally used as carboxyl activating agent for amide bonds with primary amines. EDAC has been used in peptide synthesis [96], modification of polysaccharides [15], crosslinking proteins to nucleic acids [97-98] and preparation of immunoconjugates [99].

For polysaccharides, the typical amidation procedure was described by Danishefsky [15] applied to a mucopolysaccharide. In many studies [16-18], their reaction conditions were followed: mixing of all components at pH 4.75. But, some reports claimed successful formation of amide link [15-16] whereas others reported failure in amidation [17-18]. Recently, the use of N-hydroxysuccinimide like an activating reagent was found to be effective for the EDAC mediated amidation [10, 18].

The amide bond formation between carboxylic acid functions and amines is catalyzed by carbodiimide throughout activation of carboxylate functions (Figure 17). Carbodiimide reacts with carboxyl groups to form an unstable, intermediate O-acylurea which, in the absence of nucleophiles, rearranges to a stable N-acylurea by way of a cyclic electronic displacement. From this active O-acylurea, a more hydrolysis-resistant and nonrearrangeable active ester intermediate is formed to graft primary amines to oxidized cellulose samples (Figure 17). NHS, commonly found in organic chemistry, is further used as an activating reagent. Activated acid functions (NHS-ester intermediate) react with amines to form amide bonds whereas nonactivated carboxylic acid functions would just form a salt with an amine. NHS then reacts to form a less labile activated acid and EDAC is released subsequently like water-soluble derivative of urea.

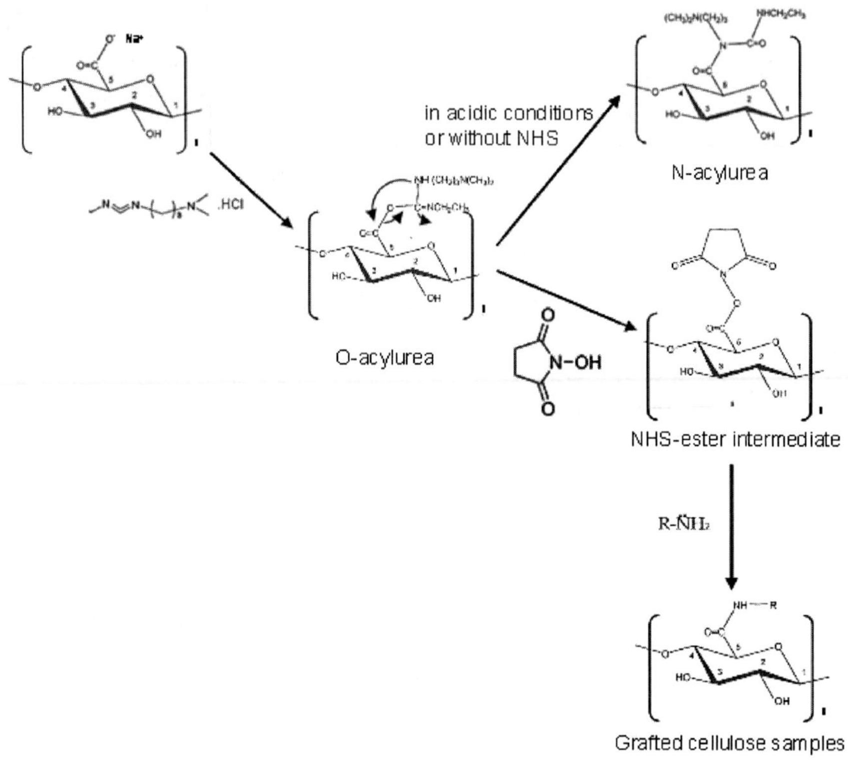

Figure 17. Grafting reaction onto oxidized cellulose samples.

3.2.) Influence of Grafting Conditions

At low pH, carbodiimides are considered as unstable [15, 17-18]. In fact, under acidic conditions, the unstable intermediate O-acylurea is readily formed and rearranges by the cyclic electronic displacement to the stable N-acylurea. Indeed, this O→N migration occurred in presence of a primary amine [17] and any amide formation occurs.

Furthermore, nucleophilic addition to the ester formed requires amine to be presented in an unprotonated form. Only a limited number of amines have pKa values in the suitable range of **Danishefsky' procedure and are** consequently few reactive with active NHS-ester intermediate. In fact, in most cases, primary amines have typically pKa higher than 9. Therewith, the NHS-ester intermediate allows for the grafting reaction to be carried out at neutral or slightly basic pH (about 7-8.5) and consequently yields products by reaction with primary amines. The combination of these observations suggested amidation under alkaline conditions. Few authors have already presented this way better than **Danishefsky' procedure** to form amide bonds onto cellulose in an alkaline medium [22, 100].

3.3.) Grafting Reaction onto Polyglucuronic Acid

For having polyglucuronic acid as starting material, we decided to use cellulose III obtained from native cellulose by ammonia treatment. Therefore, cellulose III presented crystalline structures with primary hydroxyl groups relatively accessible and therefore could be highly oxidized, leading to almost totally soluble polyglucuronic sample. The TEMPO mediated oxidation has been applied to cellulose III_I leading to complete solubilization of product. This procedure has been frequently used to improve the reactivity of crystalline cellulose for the preparation of derivatives.

The grafting reaction was carried out on polyglucuronic acid in the presence of carbodiimide and NHS with linear and cyclic amines. The conditions were 2.5 mmol amine / 1 mmol glycosyl unit (2.5 eq). A mixture of EDAC and NHS (1.5 eq each) diluted in 2 mL of water were added [22]. At stable pH 8, the mixture of EDAC and NHS was added. The solution became turbid and the pH value tending to increase is maintained to 8 by adding 0.5 M HCl and/or NaOH solutions. At the end of grafting reaction, the samples were collected. The products issued to grafting reaction were analyzed by conductimetry, elemental analysis, FTIR, NMR and EPR spectroscopies.

Degree of conversion. The degrees of conversion calculated from conductimetry, elemental analysis and EPR results from equations 3, 6, 10 are summarized in Table 8 with a significant agreement [22]. We obtained degrees of conversion close to 13%, 33%, 39% and 56% with *n*-octyl amine, 4-amino TEMPO, *n*-butyl amine and 2-methoxy ethyl amine, respectively [22]. These results indicated a decreasing dependence with the increase of carbon chain length of amines due to the variable electronic distribution along the amine chain, the steric effects and more probably to a decrease in solubility. But, these informations can not fully explain the results.

Table 8. Degrees of conversion from polyglucuronic acid.

Amines	Conductimetry	Elemental analysis	EPR analysis
n-butyl amine	0.39	0.43	-
n-octyl amine	0.13	0.15	-
2-methoxy ethyl amine	0.56	0.59	-
4-amino TEMPO	0.33	0.37	0.36

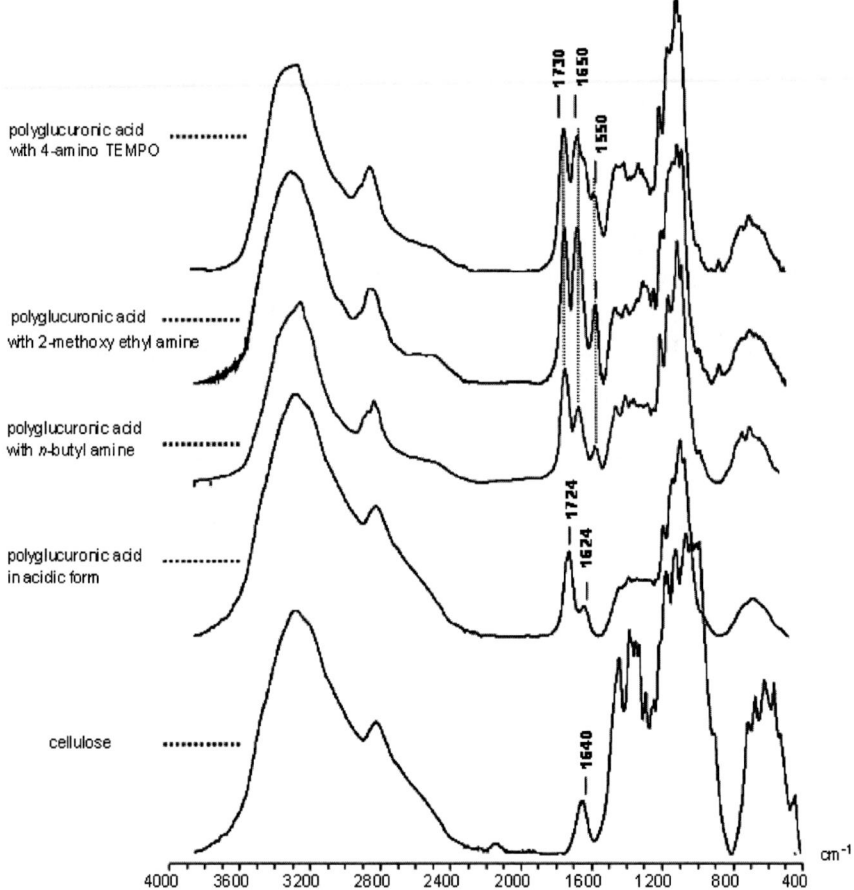

Figure 18. FTIR spectra of cellulose, polyglucuronic acid, polyglucuronic acid-*n*-butyl amine, polyglucuronic acid-2-methoxy ethyl amine and polyglucuronic acid-4-amino TEMPO.

In fact, the amine structures are characterized by inductive and mesomer effects and so the pKa values have to be into account. In aqueous medium, the protonation of amines provides ammonium salts (RNH_3^+). As during the grafting reaction the pH was maintained at around 8, the amount of ammonium ions in solution was lower for amines with pKa values nearest to 8. Thus, for instance, 2-methoxy ethyl amine (pKa=9.2) in solution was presented less ammonium ions than for *n*-butyl amine (pKa=10.6).

Since the nucleophilic addition to the NHS-ester intermediate requires the amine to be presented in an unprotonated form and carried out at slightly basic pH (as above-mentioned), then amines which will be less protonated should present a better reactivity. That's observed with 2-methoxy ethyl amine which presents the lowest pKa value and the higher degree of conversion in the range tested. From these results it appears that the grafting can be directly correlated with the amines reactivities.

FTIR spectroscopy. By FTIR spectroscopy, the products issued to grafting reaction are presented in Figure 18.

Characteristic carboxyl acid peak appears at 1730 cm^{-1} with two other distinct bands at 1650 and 1550 cm^{-1} located in the zone related to the (-CONH-), corresponding respectively to the (C=O) stretching band and to the (-NH) bending vibration band [101]. Indeed, as already reported by several authors [18, 102-103], the carboxylic acid band is well separated from the amide group (-CONH-) bands at 1650 and 1550 cm^{-1}. The FTIR analysis clearly shows that an amide bond has been formed during the grafting reaction and a correlation between degrees of conversion and the intensity of characteristic peaks is indicated.

Liquid-state NMR characterization. The chemical shift data of polyglucuronic acid-*n*-butyl amine, 2-methoxy ethyl amine and polyglucuronic acid-4-amino TEMPO are reported in Table 9 and Table 10. The characteristic signals related to carbons of polyglucuronic acid are observed, and among them, several new resonances are detected. The presence of an extra signal well separated from the (-COONa) C6 resonance is observed at 170.30 ppm with linear amines and 174.75 ppm with 4-amino TEMPO and assigned to (-CONH-) acetamide carbon according to ^{13}C NMR data already published by several authors [21, 104-105]. The presence of this signal supported the amide bond formation [22]. The other new resonances in the spectra of polyglucuronic acid grafted to amines are equally attributed. In order to view typical NMR results, the ^{13}C NMR spectrum of polyglucuronic acid-2-methoxy ethyl amine is reported in Figure 19.

Table 9. Chemical shift data of polyglucuronic acid based products with *n*-butylamine and with 2-methoxy ethyl amine. All the experimental data used have been extracted from previous work [22].

Polyglucuronic acid with *n*-butyl amine				Polyglucuronic acid with 2-methoxy ethyl amine			
^1H δ (ppm)		^{13}C δ (ppm)		^1H δ (ppm)		^{13}C δ (ppm)	
H1	4.69	C1	103.5	H1	4.44	C1	102.85
H2	3.56	C2	73.75	H2	3.36	C2	73.20
H3	3.81	C3	75.25	H3	3.51	C3	75.60
H4	3.83	C4	81.95	H4	3.60	C4	81.15
H5	4.05	C5	76.30	H5	3.81	C5	75.70
		C6 carbonyl	175.65			C6 carbonyl	175.25
		C6 acetamide	170.30			C6 acetamide	170.25
H_3C	1.08	CH_3	13.90	$O-CH_3$	3.28	$O-CH_3$	58.65
H_2C	1.51, 1.69	CH_2	20.35, 31.20,	$H_aH_b-C_1$	3.28	$O-C_2$	70.60
		-HN-C	40.25	$H_cH_d-C_2$	3.45	$-HN-C_1$	39.60

The ^1H spectra of polyglucuronic acid grafted amines were analyzed and the different proton signals assigned by 2D-COSY and 2D ^{13}C-^1H HMQC experiments. For illustration, the 2D ^{13}C-^1H HMQC spectrum of polyglucuronic acid-2-methoxy ethyl amine was presented in Figure 20a. All the characteristic proton signals of oxidized cellulose (H1, H2, H3, H4, H5) were easily identified and are in agreement with assignment already published [12, 106]. The presence of characteristic signals both in the ^1H and ^{13}C NMR spectra confirmed the grafting onto C6 carbons of polyglucuronic acid.

The 2D ^{13}C-^1H HMBC spectrum presented in Figure 20b reported strong cross-peaks between C6 acetamide (O=C-NH) and H_a and/or H_b protons of amine carbon C_1. This signal confirmed and corroborated the formation of the chemical bond between amine and carboxyl groups of the glucuronic unit. An additional strong connectivity between C6 acetamide (O=C-NH) and H_c and / or H_d protons of amine carbon C2 across 4 linkages suggesting a "w" conformation, confirmed the coupling between the carboxy group and 2-methoxy ethyl amine.

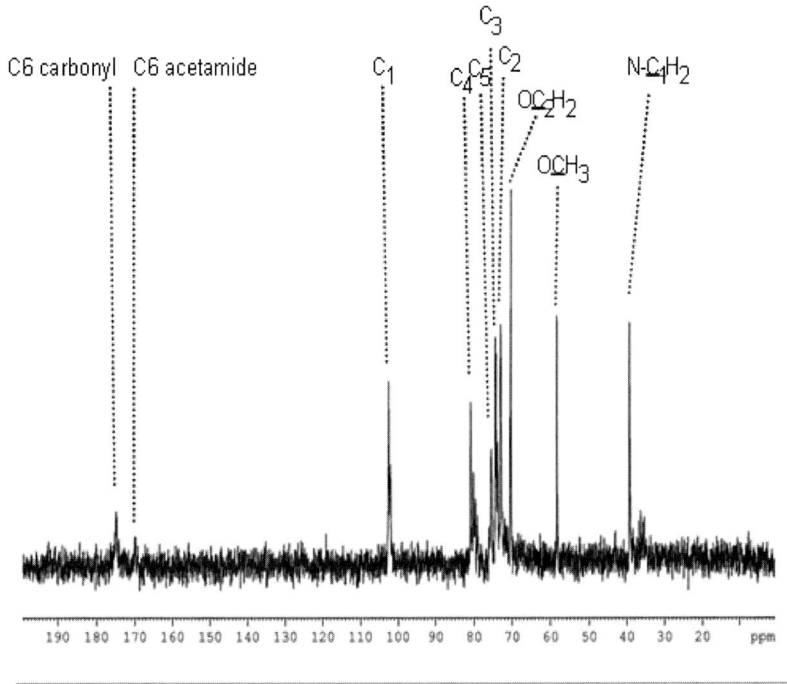

Figure 19. ^{13}C NMR spectra in D_2O of polyglucuronic acid-2-methoxy ethyl amine. All the experimental data used have been extracted from previous work [22].

For the grafting of 4-amino TEMPO, the ^{13}C NMR data are reported in Table 10. ^{13}C signals detected are attributed to the nitroxide radical, as already published by Irwin [107] with resonances at 56.40, 43.75, 39.45, 25.90, 15.35 ppm assigned to quaternary, tertiary, secondary and primary (25.90 and 15.35 ppm) carbons, respectively and the signal at 174.75 ppm to (-CONH-) acetamide C6 carbon. The evidence of coupling between polyglucuronic acid and 4-amino TEMPO was supported and highlighted by the 2D ^{13}C-^{1}H HMBC NMR experiments (Figure 21) where long range coupling constant could be observed via cross-peak correlations. The presence of a strong cross-peak between C6 of polyglucuronic acid and H$_{4'}$ of nitroxide radical is observed. This correlation corresponds to a connectivity across three bonds and clearly indicates the amide bond formation. Strong connectivities between C1 and H4* and H1 and C4* corresponding to *inter* residual three-bond correlations over the glycosidic linkages are equally observed.

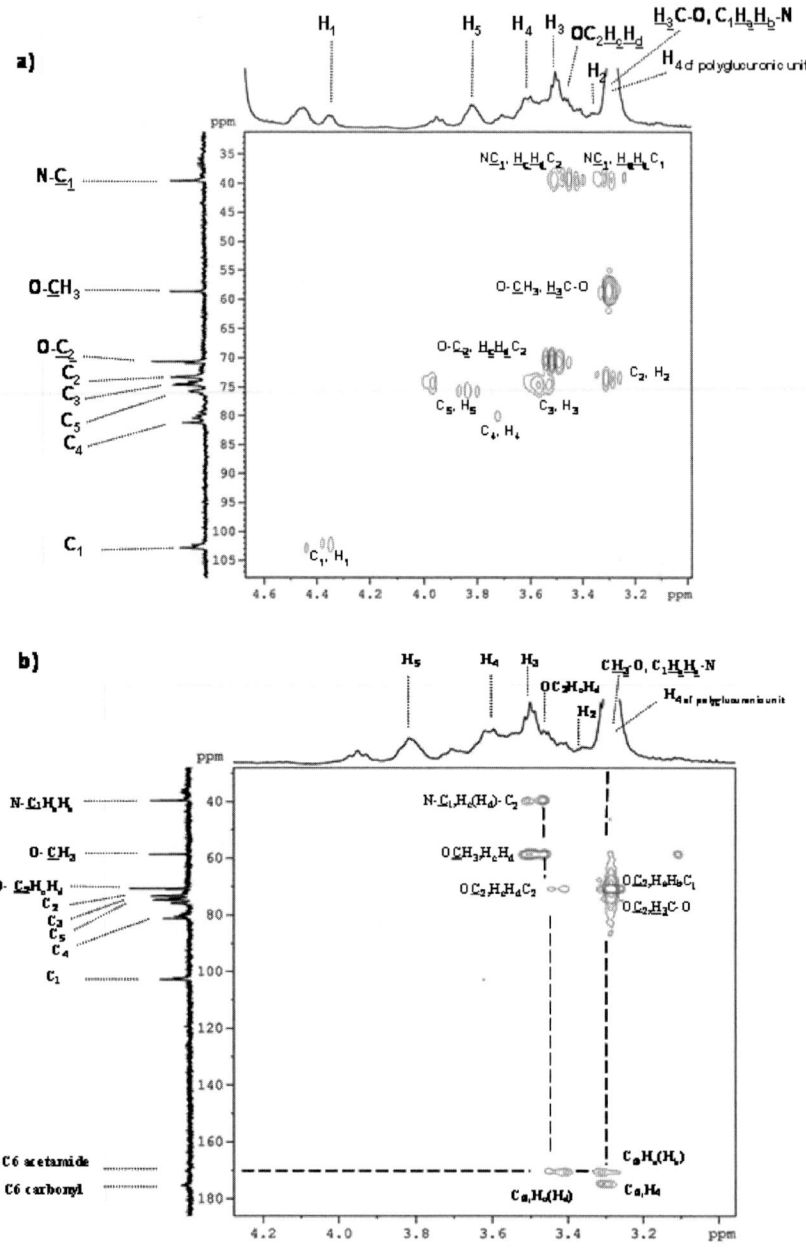

Figure 20. 2D ^{13}C-^{1}H spectra in D$_2$O: **a)** HMQC of polyglucuronic acid-2-methoxy ethyl amine; **b)** HMBC of polyglucuronic acid-2-methoxy ethyl amine. All the experimental data used have been extracted from previous work [22].

Table 10. Chemical shift data of polyglucuronic acid-4-amino TEMPO. All the experimental data used have been extracted from previous work [22].

^1H δ (ppm)		^{13}C δ (ppm)	
H1	4.69	C1	103.35
H2	3.55	C2	73.75
H3	3.82	C3	75.25
H4, H4*	3.85	C4, C4*	81.95
H5	4.04	C5	76.30
		C6 carbonyl	175.65
		C6 acetamide	174.75
H4'	2.60	CH_3	15.35, 25.90
H_2C3', H_2C5'	1.68	C4'	39.45
H_3C-C2', H_3C-C6'	1.43	C3', C5'	43.75
		C2', C6'	56.40

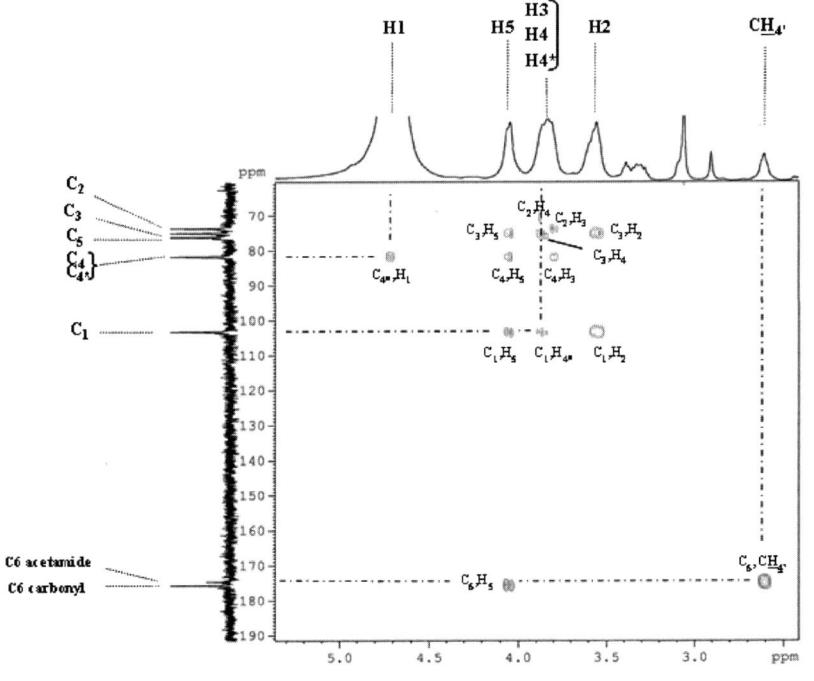

Figure 21. 2D ^{13}C-^1H HMBC spectrum in D_2O of polyglucuronic acid-4-amino TEMPO. All the experimental data used have been extracted from previous work [22].

Based on these results and observations, the chemical structure of the grafted polyglucuronic acid can be established and reported in Figure 22.

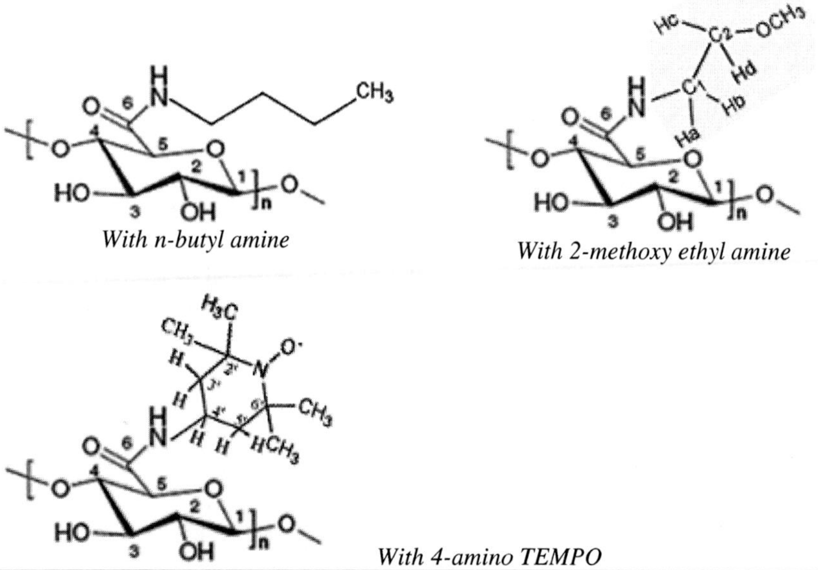

Figure 22. Chemical structure of amines grafted polyglucuronic acid.

We observed that the limitation of the grafting reaction is the result combined of amine reactivity connected to pKa value and pH solution, the presence of alkaline medium for carbodiimide mediated reaction and steric effects inherent in all molecules. The liquid-state NMR characterization probed the effectiveness of grafting reaction throughout the presence of characteristic strong cross-peaks reported in the 2D NMR experiments.

This first encouraging part has led us to develop the amines-grafting onto water-insoluble substrates of cellulose. We headed towards cellulose nanocrystals that are partially oxidized in surface.

3.4.) Grafting Reaction onto Surface Carboxylated Cellulose Nanocrystals

From native cellulose, a partial conversion into surface carboxylated cellulose nanocrystals has been realized owing to its high crystallinity and its poor accessibility of primary hydroxyl groups engaged in both *intra* and *inter* molecular hydrogen bonds in the native cellulose structure. The oxidation led to a two-phase solution: water-soluble and water-insoluble fractions. These two fractions corresponded to the totally oxidized polyglucuronic acid and the surface carboxylated cellulose nanocrystals, respectively.

The grafting reaction by carbodiimide mediated amidation process is carried out onto surface carboxylated cellulose nanocrystals with nitroxide radical (4-amino TEMPO). A suspension is obtained. Thereafter, the purification of products included precipitation, dialysis and freeze-drying steps. The reaction yields reached 98% with cotton linters. This slight loss of materials can been explained by partial degradation of grafting product. The importance of material loss depended on the crystal size of cellulose whiskers which is more important when the soluble fraction is high. The products issued to grafting reaction were analyzed by conductimetry, elemental analysis, FTIR, NMR and EPR spectroscopies.

Table 11. Degrees of conversion after reaction with 4-amino TEMPO.

Substrate	Degree of conversion (%)			
	Conductimetry	Elemental analysis	EPR analysis	NMR measurement
Surface carboxylated cotton linter cellulose nanocrystals	31	32	28	30

Degrees of conversion. The degrees of conversion of products are calculated from various characterizations. The results are summarized in Table 11. When using the conductimetry and elemental analysis and the aforementioned equations 3 and 10, there is a remarkable concordance in the values of the *DC*. The average value is 30% with cotton linters (Table 11).

FTIR spectroscopy. The Figure 23 reported the FTIR spectra of cotton linters cellulose. Characteristic strong absorption bands near 3340-3338 cm^{-1} assigned to the (OH) stretching band and near 1640 and 2900 cm^{-1} attributed to water molecules adsorbed onto the surface of carboxylated nanocrystals and (-CH) stretching vibration belonging to the anhydroglucose unit, with strong intensity, are observed.

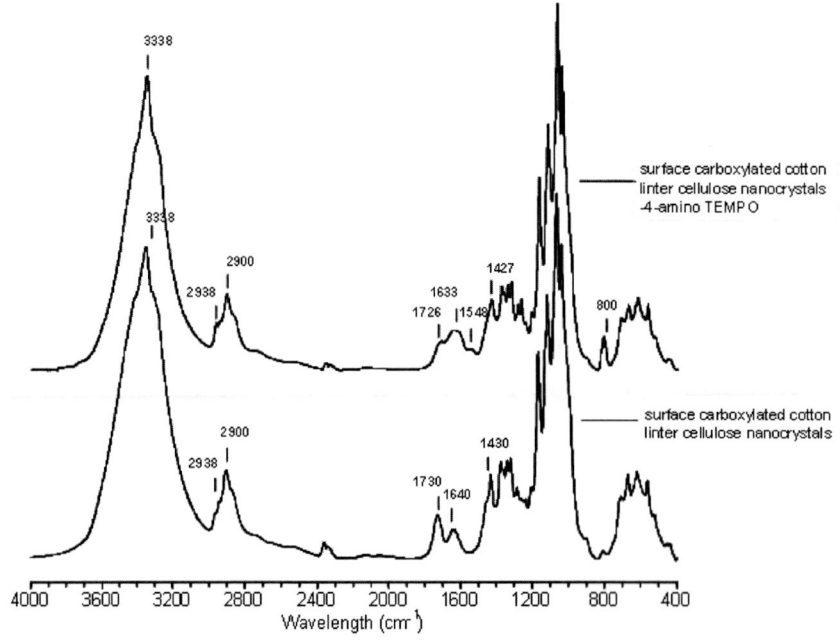

Figure 23. FTIR spectra of surface carboxylated cotton linters nanocrystals reacted with 4-amino TEMPO.

The carboxyl acid peak occurs near 1726 cm^{-1} and shows significantly reduced intensity compared with that of the unreacted cellulose nanocrystals. Two other distinct bands at 1630 and 1550 cm^{-1} are also observed, and corresponded to the (C=O) stretching band (amide I) and to the (-NH) bending vibration band (amide II), respectively [101] as above-mentioned. Indeed, the carboxylic acid peak at 1726 cm^{-1} is well separated from the amide group (-CONH-) bands, as already reported by several authors [18, 102-103]. In agreement with present results, Araki and collaborators [10] have also observed two absorption bands at 1657 and 1544 cm^{-1} corresponding to amide I and amide II absorptions resulting from the effective binding of PEG-NH$_2$ onto rodlike cellulose microcrystals. A significant reduction in the intensity of the free (COOH) stretching vibration band (1730 cm^{-1}) is also noted in comparison with the spectra of ungrafted oxidized nanocrystals confirming a reduction of the number of (-COOH) functions. These observations are similar to the polyglucuronic acid samples spectra. This result confirms the amidation occurring at the C6 carbon.

A slightly higher intensity of the 800 cm^{-1} peak assigned to (-CH-) bending vibration band (δ_{C-H}) is also noted in the spectra of the grafted sample. The presence of (-CH-) groups on the 4-amino TEMPO involved the amplitude. The FTIR results confirm the conductometric and elemental analysis data. Similar observations were made when polyglucuronic acid in solution was used as starting material [22] for grafting.

EPR Spectroscopy. Electron Paramagnetic Resonance (EPR) spectroscopy has been shown to be a reliable technique to characterize and quantify the microstructural and dynamic properties of various species.

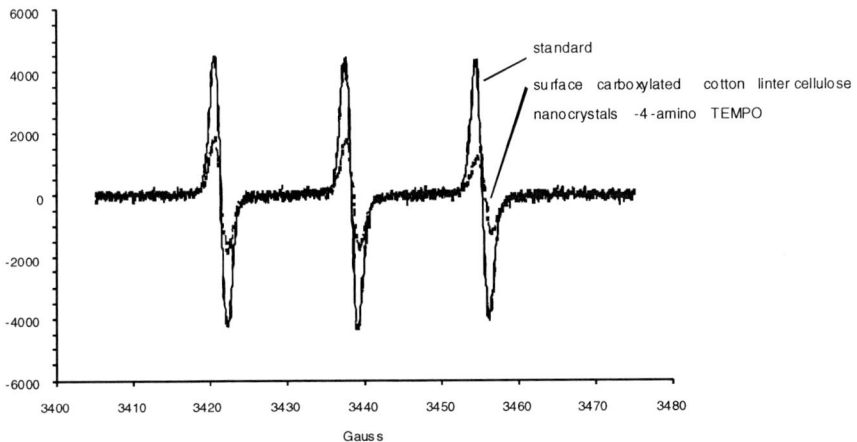

Figure 24. EPR spectra of 4-amino TEMPO (standard) and surface carboxylated cotton linter cellulose nanocrystals-4-amino TEMPO.

The EPR spectra of surface carboxylated cotton linter cellulose nanocrystals-4-amino TEMPO and an aqueous solution of 4-amino TEMPO as standard reference are presented in Figure 24. The three symmetrical well-resolved derivatives of Lorentzian lines are generally characteristic of the presence of free radical in solution. For conjugate sample, the lines broadening (particularly the lines at high field) clearly show a restrictive motion of the spin label linked to the cotton linters cellulose nanocrystals. It is therefore an indication that the nitroxide moiety has been successfully grafted on the nanocrystals. By double integration (area) of the labeled nanocrystals spectrum, the absolute number of grafted radicals is obtained (Table 11) and the degree of conversion calculated (equation 9) is close to those obtained with others characterizations.

Solid-state NMR characterization. During the past decade, high resolution solid-state NMR spectroscopy has proven to be a powerful tool in the investigation of structural features of cellulose materials [35, 95, 108]. The ^{13}C NMR spectra of cellulose raw materials presented characteristic signals in agreement with the literature: resonances at 62.4 and 65 ppm assigned to disordered and crystalline regions of C6 carbons of cellulose and between 70-77 ppm in the form of two very intense peaks the resonances assigned to C2, C3, and C5 carbons. The signals at 83.5 and 88.7 ppm signals are attributed to disordered and crystalline regions of C4 carbons, respectively, and C1 carbons are located at 105 ppm [9, 35, 95, 109]. The resulting cellulose nanocrystals samples were characterized by solid-state NMR spectroscopy. The Figure 25 illustrated the CP-MAS ^{13}C NMR spectrum after oxidation and amidation. After oxidation, the major change concerns the appearance of the carboxyl groups signal at 174.8 ppm. The *Do* is evaluated by the integration of the signal at 174.8 ppm which displays high amplitude. A crystallinity index [35] was estimated by comparing the surface area of the C4 signals at 83.5 and 88.7 ppm corresponding to the carbons in disordered and crystalline regions, respectively.

After grafting by amidation, we observed the presence of characteristic carbons signals at 40 ppm corresponding to the C3', C5' and C4' of 4-amino TEMPO, the notable reduction of resonance due to the disordered regions of C6 carbons (62.4 ppm), the strong reduction in intensity of the signal at 83.5 ppm attributed to disordered regions of C4 carbons, and the decrease of the carboxyl signal of carboxylated nanocrystals at 174.8 ppm. Indeed, the C5 carbons resonance at 77 ppm presents a reduced intensity revealing a change in spatial environment due to the grafting to C6 carbons. In addition, a new resonance is detected at 171.5 ppm well separated from the (-COONa) C6 resonance (174.8 ppm). Based on the aforementioned results [19, 22, 104, 107], this new resonance corresponded to (-CONH-) acetamide carbon resulting from the grafting of 4-amino TEMPO on carboxylated nanocrystals. Irwin and collaborators [107] have also observed resonances at 40 ppm assigned to C3', C5' and C4' of 4-amino TEMPO during the preparation of the amidated poly(galacturonic acid).

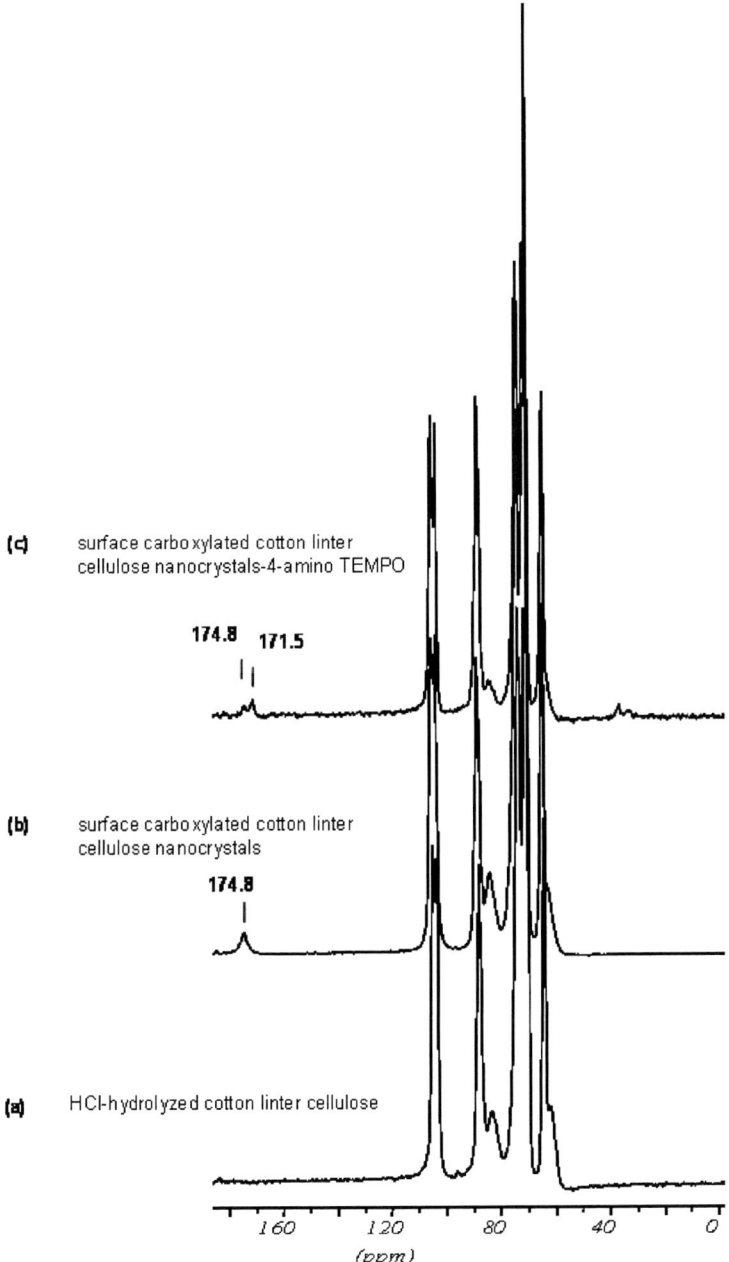

Figure 25. CP-MAS ^{13}C NMR spectra of (a) cotton linter samples, (b) surface carboxylated cotton linter cellulose nanocrystals and (c) surface carboxylated cotton linter cellulose nanocrystals-4-amino TEMPO.

Table 12. Results of the quantitative analysis of CP-MAS ^{13}C NMR spectra.

Cellulose samples	NaClO C=O[b] (molar ratio)[a]	C=O[b] (174.8 ppm)	C6cryst[c] (68-63 ppm)	C6amor[c] (63-58 ppm)	C6total[d] (68-58 ppm)	C4cryst[c] (CI[e]) (91-86 ppm)	C4amor[c] (86-81 ppm)
cotton linter cellulose	0	0	0.54	0.37	0.92	0.69	0.33
hydrolyzed Cotton linter cellulose	0	0	0.68	0.22	0.90	0.75	0.25
surface carboxylated cotton linter cellulose nanocrystals	1	0.10	0.68	0.13	0.81	0.72	0.28
surface carboxylated cotton linter cellulose nanocrystals-4-amino TEMPO	1	0.07	0.68	0.09	0.77	0.72	0.19

[a] Mol NaClO/mol glycosyl unit. [b] Carboxyl groups (D_o) obtained by integration of the signal at 174.8 ppm. [c] Crystalline and amorphous contents determined by deconvolution. [d] Obtained by integration of the signal at 62.4 ppm. [e] Crystallinity index.

The Table 12 reveals that the carboxyl group content decreases with the grafting of 4-amino TEMPO concomitant with the reduction of the disordered regions of C6 carbons resonance at 62.4 ppm, whereas the crystalline contribution at 65 ppm remains constant. This is a strong indication of the selectivity of reaction which occurs only at the surface of nanocrystals and / or disordered regions. The combination of this reduction and the presence of the acetamide resonance at 171.5 ppm confirm that the grafting occurred in the disordered regions of C6 carbons. The decrease of the C4 signal at 83.5 ppm (disordered regions) corroborates it. Furthermore, it is worth noting that the crystallinity index does not evolve during the grafting reaction (Table 12). This observation is a strong indication that the integrity of the crystallites is not altered by the amidation.

In addition, the presence of acetamide carbon on the conjugate induce modifications in environment of C4 and C6 carbons in term of change in disorder, which are observed by a reduction in the intensity of corresponding resonances at 62.4 and 83.5 ppm. This phenomenon is attributed to the amine-grafting onto surface of nanocrystals.

4.) DISCUSSION ON THE REACTIVITY OF GRAFTED CELLULOSE SUBSTRATES

Polyglucuronic acid. The main restriction of amine-grafting on totally oxidized cellulose is focused on the protonation state of amines during the carbodiimide mediated amidation accentuated by the decrease of solubility with carbon chain length increase. The pKa value must be taken into account. Higher the pKa value of amine is, smaller the grafting is. The use of amines presenting pKa values close to 9 would be highly relevant for grafting mediated by EDAC in aqueous conditions. As a result of this less protonated state of molecule, amines present a better reactivity.

In addition, the length of the amine carbon chains involved steric effects that minimize the amide bond formation onto C6 carbons of oxidized cellulose. A similar response is observed with large amine like 4-amino TEMPO. To knowledge of authors, few reports presented conversion oxidized C6 carbons to acetamide carbons. The conversion is thus driven by the ability of amines to be protonated in alkaline conditions.

Surface carboxylated cellulose nanocrystals. The average *DC* value for the cotton linter cellulose is near 30%. In fact, a maximum of about one third of carboxylated functions of cellulose are really grafted. Similar results are reported in the literature, for instance, with the grafting of single terminally aminated PEG (20%) or phenylpropanolamine (6%) onto water-insoluble oxidized cellulose [10, 19]. For attempting to understand the results, the organization of cellulose in plant cell walls and the organization of cellulose nanocrystals resulting from oxidation must be considered.

According to current models and recent works [49, 95, 110-112], the cellulose microfibrils present square section with exposed surfaces corresponding to the $[1\overline{1}0]$ and $[110]$ crystal planes of the cellulose lattice (in reference to the cellulose unit cell defined by Sugiyama [49]). As opposed to the core of the microfibrils, which is considered to be in a crystalline arrangement, the microfibril surfaces are organized in a different manner. The difference between surface and core chains is clearly evidenced in CP/MAS ^{13}C NMR spectra of cellulose: different resonances at C4 carbons are attributed either to the core or surface cellulose chains [35, 113]. One distinguishes surfaces accessible to oxidation located at the exterior of microfibrils aggregates and inaccessible surfaces located inside the microfibrils aggregates. This organization in aggregates is not modifying by HCl hydrolysis that only converts microfibrils aggregates into nanocrystals aggregates. In addition, regarding the description of the [110] and [110] crystal surfaces, the oxidation took place only one-half of the surface hydroxymethyls of cellulose and the other half being toward the core of the crystalline domains (Figure 26).

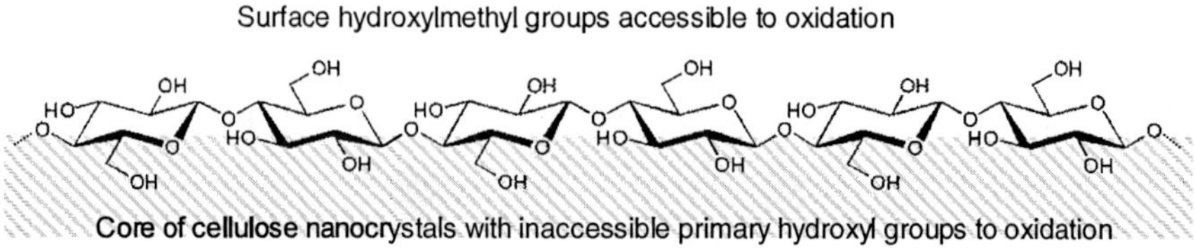

Surface hydroxylmethyl groups accessible to oxidation

Core of cellulose nanocrystals with inaccessible primary hydroxyl groups to oxidation

Figure 26. Scheme of cellulose chain at the surface of cellulose crystal.

Theoretically, if all surface hydroxylmethyl groups were transformed during carboxylation, the maximum grafting would be 50%. The experimental degrees of conversion opposed to the theory.

It is worth interesting to make some comments on the experimental results of grafting. In first observation, even if the HCl hydrolysis broadly involved the production of nanocrystals exempt to amorphous domains, crystals aggregates are obtained. Without HCl hydrolysis, the crystals are more highly aggregated [95]. After carboxylation, the maximal Do of cellulose substrates corresponded to 15%. Even if the surface oxidation improved the individualization of the crystals, less one-half of the surface hydroxymethyls of cellulose is accessible. A part of these accessible surfaces is entangled inside cellulose aggregates throughout hydrogen bonds network between nanocrystals embedded in aggregate. The oxidized samples consist rather of crystallites made of two or three individual crystals. On the basis of these informations, the 50% of conversion is not reachable.

Further, the partial individualization of the crystals achieved a modification of accessibility rate of carboxylated functions for grafting since the aggregates made of two or three individual crystals are organized in different and inhomogeneous manner. We can envisage a large scale of arrangements from a packing of two or three individual crystals (parallel until displacement across of nanocrystals). The least favorable cases for grafting reaction are presented in Table 13 in comparing with nanocrystals totally individualized with 50% conversion. We can consider that the limit of DC values corresponds to 33% up to 50%. By and large, the new calculated DC value resulting from the least favorable case with three nanocrystals by aggregates is very close to the experimental results exhibited in this work.

Presumably, the steric effects resulting to the amine grafted adding to bulk organizations of aggregates could partly explain the experimental results.

Table 13. Schematic representation of most limiting cellulose nanocrystals organizations for the amine-grafting reaction.

	Cellulose organization	nanocrystals	Accessibility	DC calculated
a)	nanocrystal individualized	totally	100%	50%
b)	cellulose nanocrystals made of three individual crystals with total overlapping	A B C	A: 75% B: 50% C: 75%	33,3%
c)	cellulose nanocrystals made of two individual crystals with total overlapping	A B	A: 75% B: 75%	37,5%

CONCLUSION

In this work, we described the grafting on polyglucuronic acid samples and surface carboxylated cellulose nanocrystals with single terminally aminated molecules. To achieved it, we used linear and cyclic amines presenting pKa values at around 9 – 11. Polyglucuronic acid was issued by TEMPO mediated carboxylation of cellulose III_I. Surface carboxylated cellulose nanocrystals were obtained by oxidation of HCl-hydrolyzed cellulose in order to realize a preferential surface oxidation leading to higher nanocrystals yields. The products resulting from grafting reaction were analyzed by conductimetry, elemental analysis, FTIR, NMR and EPR spectroscopies.

After carboxylation of cellulose by TEMPO mediated oxidation, the main observations corresponded to the appearance of the carboxyl groups signals near 175 ppm with the carboxyl peak at 1730 cm^{-1}. The NMR characterization of cellulose substrates indicated that oxidation is selectively realized on C6 carbons.

Upon amidation, a covalently-linked cellulose with a 1:2,5:1,5 mol ratio of cellulose:amine:carbodiimide is formed. We demonstrate that, by the formation of a more hydrolysis-resistant and nonrearrangeble active ester intermediate from the active O-acylurea, the primary amine-grafting onto oxidized cellulose is possible. With polyglucuronic acid, the reaction is driven by protonation state and solubility of amines in the alkaline conditions used. With surface carboxylated cellulose nanocrystals, the results widely vary from the surface oxidation level and the morphology of carboxylated samples, i.e. nanocrystal aggregates made of two or three individual nanocrystals. By NMR characterization, the presence of acetamide signal is proven accentuated by the observation of two FTIR peaks at 1650 and 1550 cm^{-1} corresponding to the

(C=O) stretching band and to the (-NH) bending vibration band of (-CONH-) functions. For the grafting of 4-amino TEMPO, EPR technique equally allowed to ensure the amide bond. The combination of NMR, EPR spectroscopies, conductimetry and elemental analysis providing the undisputed evidence that the amine-grafting reaction is realized with a good agreement between calculated results.

Through this work, some parameters are identified as key drivers to success. In order to validate the experimental results, some possibilities are suggested in open discussion concerning the reactivity of cellulose substrates. With totally oxidized cellulose such as polyglucuronic acid, the grafting reaction is limited by the molecular weight of primary amines, their solubilities at reactional temperature and especially the unprotonated state of amine at the pH used. Indeed, on the basis of reactional mechanism, the active ester intermediate must obviously react with amine groups of molecule.

With partially oxidized cellulose like nanocrystals, the reaction is more dependent on substrate and particularly on the accessibility of reactive functions that are carboxylated groups. The cellulose chains organization limited the number of reactive functions to one-half of the surface hydroxymethyls of cellulose preventing partly the efficiency of amidation to one-third grafting.

ACKNOWLEDGMENT

N. FOLLAIN is grateful to CNRS for financial support during post-doctoral position at CERMAV. The author thanks Mrs M-F MARAIS (CERMAV, CNRS) for preparing the cellulose samples. The author also acknowledges the help of Dr Serge GAMBARELLI (CEA-Grenoble, DRFMC/SCIB) for the EPR characterization, M. TRIEWEILER for NMR data (CERMAV, CNRS), Dr Michel VIGNON, Dr Suzelei MONTANARI and Dr. Henri CHANZY (CERMAV, CNRS) for valuable discussion.

REFERENCES

[1] Chanzy, H. Aspects of cellulose structure. In: Kennedy JF, Phillips GO, Williams PA (eds) *Cellulose sources and exploitations*. Ellis Horwood Ltd, NY, 1990; pp 3-12.

[2] Orts, W. J.; Godbout, L.; Marchessault, R. H.; Revol, J.-F. *Macromolecules* 1998, *31*, 5717-5725.

[3] de Nooy, A. E.; Besemer, A. C.; van Bekkum, H. *Carbohydr. Res.,* 1995, *69*(1), 89-98.

[4] Bragd P.L.; Van Bekkum, H.; Besemer, A.C. *Top. Catal.,* 2004, *27*, 49-66.

[5] de Nooy, A. E.; Besemer, A. C.; van Bekkum, H. *Recueil des Travaux Chimiques des Pays-Bas,* 1994, *113*(3), 165-166.

[6] de Nooy, A. E.; Besemer, A. C.; van Bekkum, H.; van Dijk, J. A. P. P.; Smit, J. A. M. *Macromolecules*, 1996, *29*(20), 6541-6547.

[7] Chang, P. S.; Robyt, J. F. *Journal of Carbohydrate Chemistry*, 1996, *15*, 819-830.

[8] Sierakowski, M. R.; Milas, M.; Desbrières, J.; Rinaudo, M. *Carbohydr. Polym.*, 2000, *42*(1), 51-57.

[9] Isogai, A.; Kato, R. H. *Cellulose* (London), 1998, *5*(3), 153-164.

[10] Araki, J.; Wada, M.; Kuga, S. *Langmuir,* 2001, *17*(1), 21-27.

[11] Muzzarelli, R. A. A.; Muzzarelli, C.; Cosani, A.; Terbojevich, M. *Carbohydr. Polym.*, 1999, *39*(4), 361-367.

[12] Tahiri, C.; Vignon, M. R. *Cellulose* (Dordrecht, Netherlands), 2000, *7*(2), 177-188.

[13] Da Silva Perez, D.; Montanari, S.; Vignon, M. R. *Biomacromolecules,* 2003, 4, 1417-1425.

[14] Davis, W. E.; Barry, A. J.; Peterson, F. C.; King, A. J. *J. Am. Chem. Soc.* 1943, *65*, 1294-1299.

[15] Danishefsky, I.; Siskovic, E. *Carbohydr. Res.*, 1971, *16*, 199-205.
[16] Ploehn, H.J.; Goodwin, J.W. *Faraday Discuss Chem Soc*, 1991, *77-90*, 163-167.
[17] Kuo, J.W.; Swann, D.A.; Prestwich, G.D. *Bioconjugate chemistry*, 1991, 2(4), 232-241.
[18] Bulpitt, P.; Aeschlimann, D. *J. Biomed. Mater. Res.* 1999, *47(2)*, 152-169.
[19] Zhu, L.; Kumar, V.; Banker, G. S. *Int. J. Pharm.* 2001, *223*, 35-47.
[20] Lillo, L. E.; Matsuhiro, B. *Carbohydr. Polym.*, 2003, *51*, 317-325.
[21] Novak, L.; Banyai, I.; Fleischer-Radu, J. E.; Borbely, J. *Biomacromolecules*, 2007, *8*, 1624-1632.
[22] Follain, N.; Montanari, S.; Jeacomine, I.; Gambarelli, S.; Vignon, M. *Carbohydr. Polym.*, 2008, *74*, 333-343.
[23] Ashton, W.H. US Patent 3,364,200, 1968.
[24] Barry, A. J.; Peterson, F. C.; King, A. J. *J. Am. Chem. Soc.* 1936, *58*, 333-337.
[25] Klemm, D.; Philipp, B.; Heinze, T.; Heinze, U.; Wagenknecht, W. In *Comprehensive Cellulose Chemistry*, Wiley-VCH: Weinheim, New York, Chichester, Brisbane, Singapore, Toronto, 1998; Vol I, p 152.
[26] Klenkova, N. I. *Zh. Prikl. Khim.* 1967, *40(110)*, 2191-2208.
[27] Chanzy, H.; Henrissat, B.; Vuong, R.; Revol, J.-F. *Holzforshung*, 1986, *40*, 25-30.
[28] Wada, M.; Heux, L.; Isogai, A.; Nishiyama, Y.; Chanzy, H.; Sugiyama, J. *Macromolecules*, 2001, *34*, 1237-1243.
[29] Chanzy, H.; Henrissat, B.; Vincendon, M.; Tanner, S. F.; Belton, P. S. *Carbohydr. Res.*, 1987, *160*, 1-11.
[30] Henrissat, B.; Marchessault, R. H.; Taylor, M. G.; Chanzy, H. *Polymer Commun.* 1987, *28(4)*, 113-115.
[31] Karstens, T. German patent 1 011 061, 1995.
[32] Karstens, T.; Stein, A.; Steinmeier, H. French patent 2 762 603, 1998.
[33] Holtzapple, M. T.; Jun, J. H.; Ashok, G.; Patibandla, S. L.; Dale, B. E. *Applied Biochemistry and Biotechnology*, 1991, *28-29*, 59-74.
[34] Ferrer, A.; Byers, F. M.; Sulbaran-de-Ferrer, B.; Dale, B. E.; Aiello, C. *Applied Biochemistry and Biotechnology*, 2000, *84-86*, 163-179.
[35] Heux, L.; Dinand, E.; Vignon, M. R. *Carbohydr. Polym.*, 1999, 40, 115-124.
[36] Payen, A. *Compt. Rend.; Mémoire sur la composition du tissue propre des plantes et du ligneux*; 1838, *7*, 1052-1056.

[37] Willstätter, R. *Ber.; Zur Kenntnis der Hydrolyse von cellulose I.*; 1913, *46*, 2401-2412.

[38] Crawford, R. L. *Lignin biodegradation*

[39] Raymond, Y. *Cellulose structure modification and hydrolysis*. New York: Wiley, 1986.

[40] Gardner, K. H.; Blackwel, J. *Biopolymers*, 1974, *13*(10), 1975-2001.

[41] Chu, S.S.C.; Jeffrey, G.A. *Acta Cryst.*, 1968, *24*, 830-838.

[42] Fengel, D. *Holzforschung*, 1992, *46*, 283-288.

[43] Howsmon, J. A.; Sisson, W. A. In *Cellulose and Cellulose Derivatives*, Part I, 1963, V, 231-346.

[44] Atalla, R. H.; VanderHart, D. L. *Science*, 1984, *223*, 283-285.

[45] VanderHart, D. L.; Atalla, R. H. *Macromolecules*, 1984, *17*, 1465-1472.

[46] Kolpak, F.J.; Weih, M.; Bleckwell, J. *Polymer*, 1978, *19*, 123-131.

[47] Chanzy, H.; Imada, K.; Vuong, R. *Protoplasma*, 1978, *94*, 299-306.

[48] Chanzy, H.; Imada, K.; Mollard, A.;Vuong, R.; Barnoud, F. *Protoplasma*, 1979, *100*, 303-316.

[49] Sugiyama, J.; Vuong, R.; Chanzy. H. *Macromolecules,* 1991, *24*(14), 4168-4175.

[50] Vietor, R.J.; Mazeau, K.; Lakin, M.; Perez, S. *Biopolymers,* 2000, *54*(5) 342-354.

[51] Belton, P. S.; Tanner, S. F.; Cartier, N.; Chanzy, H. *Macromolecules*, 1989, *22*, 1615-1617.

[52] Kono, H.; Yunoki, S. Shikano, T.; Fujiwara, M.; Erata, T.; Takai, M. *J. Am. Chem. Soc.*, 2002, *124*(25), 7506-7511.

[53] Nishiyama, Y.; Sugiyama, J.; Chanzy, H.; Langan, P. *J. Am. Chem. Soc.* , 2003, *125*(47), 14300-14306.

[54] Nishiyama, Y.; Langan, P.; Chanzy, H. *J. Am. Chem. Soc.*, 2002, *124*(31), 9074-9082.

[55] Sugiyama, J.; Persson, J.; Chanzy, H. *Macromolecules*, 1991, *24*(9), 2461-2466.

[56] Debzi, E. M.; Chanzy, H.; Sugiyama, J.; Tekely, P.; Excoffier, G. *Macromolecules*, 1991, *24*(26), 6816-6822.

[57] Horii, F.; Yamamoto, H.; Kitamaru, R.; Tanahashi, M.; Higuchi, T. *Macromolecules,* 1987, *20*, 2946-2949.

[58] Wada, M.; Okano, T.; Sugiyama, J. *Cellulose*, 1997, *4*(3), 221-232.

[59] Yamamoto, H.; Horii, F. *Macromolecules*, 1993, *26*(6), 1313-1317.

[60] Sugiyama, J.; Okano, T.; Yamamoto, H.; Horii, F. *Macromolecules*, 1990, *23*(12), 3196-3198.

[61] Roche, E.; Chanzy, H. *Int. J. Biol. Macromolecules*, 1981, *3(3)*, 201-206.
[62] Chanzy, H.; Henrissat, B.; Vuong, R.; Revol, J. F. *Holzforschung* 1986, *40*, 25-30.
[63] Sugiyama, J.; Okano, T. Cellulose and Wood: Chemistry and Technology, Proceedings of the tenth Cellulose Conference, 1989, 119-127.
[64] Preston, R. D.; Nicolai, E.; Reed, R.; Mallard, A. *Nature*, 1948, *162*, 665-667.
[65] Frey-Wyssling, A.; Miihlethaler, K.; Wyckoff, R. W. G. *Experientia, Mikrofibrillenbau der pflanzlichen Zellwände*, 1948, *4*, 475-476.
[66] Preston, R. D. *Physics Reports*, 1975, *21*, 183-226.
[67] Revol, J. F. *J. Mat. Sci. Letters*, 1985, 4, 1347-1349.
[68] Sugiyama, J.; Harada, H.; Fujiyoshi, Y.; Uyeda, N., *Mokuzai Gakkaishi*, 1985, *30*, 98-99.
[69] Sugiyama, J.; Harada, H.; Fujiyoshi, Y.; Uyeda, N., *Mokuzai Gakkaishi*, 1985, *31*, 61-67.
[70] Sugiyama, J.; Harada, H.; Fujiyoshi, Y.; Uyeda, N. *Planta*, 1985, *166*, 161-168.
[71] Kuga, S.; Brown, R. M. Jr. *Polymer Commun.*, 1987, *28*, 311-314.
[72] Kuga, S.; Brown, R. M. Jr. *Carbohydr. Res.*, 1988, 180, 345-350.
[73] Marchessault, R. H. *Cellulose and Wood-Chemistry and technology*, Cellulosics as advanced materials, 1989, 1-20.
[74] Chanzy, H.; Henrissat, B. *FEBS Lett.*, 1985, *184*, 285-288.
[75] Koyama, M.; Helbert, W.; Imai, T.; Sugiyama, J.; Henrissat B. *Proc. Natl. Acad. Sci. USA*, 1997, *94*, 9091-9095.
[76] Sugiyama, J.; Chanzy, H.; Revol, J.F. *Planta*, 1994, *193*, 206-265.
[77] Revol, J.F.; Godbout, L.; Dong, X. M.; Gray, D.G.; Chanzy, H. *Liquid Cryst*, 1994, *16*(1), 127-134.
[78] Revol, J.F.; Bradford, H.; Giasson, J.; Marchessault, R.H.; Gray, D.G. *Int.J. Biol. Macromol.*, 1992, *14*(3) 170-172.
[79] Azizi, S.; Alloin, F.; Dufresne, A. *Biomacromolecules*, 2005, *6*(2), 612-626.
[80] Araki, J.; Wada, M.; Kuga, S.; Okano, T. *Colloids Surf. A* 1998, *142*, 75-82.
[81] Radley, J.A. *Starch and its derivatives*, Chapter 11, Chapman and hall, London 1968
[82] Nieuwenhuizen, M.; Kieboom, A.P.G.; Van Bekkum, H. *Starch*, 1985, *37*, 192-200.

[83] Floor, M.; Kieboom, A.P.G.; Van Bekkum, H. *Recl. Trav. Chim. Pays-Bas*, 1989, *108*, 384-392.
[84] Yackel, E. A.; Kenyon, W. O. *J. Am. Chem. Soc.* 1942, *64*, 121-122.
[85] Painter T.J. *Carbohydr. Res.,* 1977, 55, 95-103.
[86] Painter, T.J.; Cesaro, A.; Delben, F.; Paoletti, S. *Carbohydr. Res.,* 1985, 140, 61-68.
[87] Maurer, K.; Reiff, G. *J. Makromol. Chem.,* 1943, *1*, 27-34.
[88] Semmelhack, M.F.; Schmid, C.R.; Cortés, D.A.; Chou, C.S.*J. Am. Chem. Soc.*, 1984, *106*, 3374-3376.
[89] Semmelhack, M.F.; Cortés, D.A. *J. Am. Chem. Soc.*, 1983, *105*, 4492-4494.
[90] Davis, N.J.; Flitsch, S.L. *Tetrahedron Letters,* 1993, *34*, 1181-1184.
[91] Besemer, A.C.; Van Bekkum, H. *Starch/Stärke,* 1994, *46*(3), 101-106.
[92] Besemer, A.C.; Van Bekkum, H. *Starch/Stärke,* 1994, *46*(3), 95-101.
[93] Herrmann, W.A.; Zoller J.P.; Fischer, R.W. *J. Organomet. Chem.*, 1999, *579*, 404-407.
[94] Gomez-Bujedo, S.; Fleury, E.; & Vignon, M. R. *Biomacromolecules*, 2004, *5*(2), 565-571.
[95] Montanari, S.; Roumani, M.; Heux, L.; Vignon, M. R. *Macromolecules,* 2005, 38, 1665-1671.
[96] Wang, D.; Li, L.; Zhang, P. *Chemical, Biological, Agricultural, Medical and Earth Sciences,* 1987, *30*, 449-459.
[97] Hoare, D.G.; Koshland, D.E. Jr. *Journal of Biological Chemistry*, 1967, *242*, 2447-2453.
[98] Thomas, J. O. *J. Mol. Biol.,* 1976, *123*, 149.
[99] Drabick, J.J. *Antimicrob. Agents Chemother.*, 1998, *42*, 583-588.
[100] Araki, J.; Kuga, S.; Magoshi, J. *J. of App. Polymer Sci.*, 2002, *85*, 1349-1352.
[101] Williams, D. H.; Fleming, I. In *Spectroscopic methods in Organic Chemistry*. McGraw-Hill, Maindenhead, Berkshire, 1966.
[102] Hu, F-Q.; Zhao, M-D.; Yuan, H.; You, J.; Du, Y-Z.; Zeng, S. *International Journal of Pharmaceutics,* 2006, *315*, 158-166.
[103] Deng, Y.; Liu, D.; Du, G.; Li, X.; Chen, J. *Polymer International,* 2007, *56*(6), 738-745.
[104] Jiang, B.; Drouet, E.; Milas, M.; Rinaudo, M. *Carbohydr. Res.,* 2000, *327*, 455-461.
[105] Toffey, A., Samaranayake, G., Frazier, C. E., Glasser, W. G. *Journal of Applied Polymer Science*, 1996, *60*, 75-85.

[106] Heyraud, A.; Courtois, J.; Dantas, L.; Colin-Morel, Ph.; Courtois, B. *Carbohydr. Res.,* 1993, *240*, 71-78.
[107] Irwin, P.L.; Sevilla, M.D.; Osman, S.F. *Macromolecules,* 1987, 20, 1222-1227.
[108] Dinand, E.; Vignon, M. R. *Carbohydr. Res.,* 2001, 330, 285-288.
[109] Wada, M.; Heux, L.; Isogai, A.; Nishiyama, Y.; Chanzy, H.; Sugiyama, J. *Macromolecules*, 2001, 34(5), 1237-1243.
[110] Preston, R.D. *The Physical Biology of Plant Cell Walls*; Chapman & Hall: London, 1974.
[111] Roelofsen, P.A. *The Plant Cell-Wall;* Gebrüder Borntraeger: Berlin-Nikolasee, 1959.
[112] Revol, J-F. *Carbohydr. Polym.,* 1982, 2, 123-134.
[113] Atalla, R.H.; VanderHart, D.L. *Solid State Nucl. Mag. Reson.* 1999, 15, 1-19.

INDEX

A

absorption, 45, 46
accessibility, 2, 7, 34, 44, 52, 56
acetic acid, 31
acid, 1, 8, 9, 11, 12, 15, 26, 27, 28, 31, 34, 35, 37, 38, 39, 40, 41, 42, 43, 44, 45, 46, 47, 48, 51, 55, 56
acidic, 10, 36
activation, 35
adhesion, vii, 2
agent, 35
agents, 29
aggregates, 27, 51, 52, 55
alcohol, vii, 28, 32
alcohols, 28, 32
aldehydes, 28, 31
algae, 15, 16, 21
alkaline, 20, 23, 30, 37, 44, 51, 55
amide, viii, 2, 35, 36, 37, 39, 41, 46, 51, 56
amine, 8, 28, 35, 36, 37, 38, 39, 40, 41, 42, 44, 50, 51, 53, 55, 56
amines, viii, 2, 6, 7, 8, 10, 35, 36, 37, 39, 40, 44, 51, 55, 56
amino, 6, 11, 12, 13, 37, 38, 39, 41, 45, 46, 47, 48, 49, 50, 51, 56
ammonia, 2, 7, 21, 37
ammonium, 39
ammonium salts, 39
amorphous, 1, 7, 20, 24, 25, 26, 34, 50, 52
amplitude, 11, 47, 48
amylopectin, 1, 30
animals, vii, 15, 16, 21
annealing, 22
application, 24
aqueous solution, 47
ascidians, 16
assignment, 22, 40

B

bacteria, vii, 15, 16, 21
bacterial, 1, 21
barrier, 16
beams, 19
bending, 39, 46, 47, 56
binding, 46
biocompatible, vii, 3
biodegradability, 31
biodegradable, vii, 3
biodegradation, 59
biofilms, 15
biopolymers, 59
bonds, viii, 10, 18, 19, 23, 24, 34, 35, 37, 41, 45, 52
branching, 18

C

capillary, 12, 13
carbohydrates, 28
carbon, 17, 21, 37, 39, 40, 41, 46, 48, 50, 51
carboxyl, viii, 2, 9, 10, 27, 31, 34, 35, 39, 40, 46, 48, 50, 55
carboxyl groups, viii, 2, 9, 10, 27, 31, 35, 40, 48, 55
carboxylic, 28, 31, 32, 35, 39, 46
carboxylic acids, 32
carrier, vii, 3
catalyst, vii, 1, 31
catalytic system, 30
CEA, 56
cell, 1, 15, 21, 22, 23, 26, 51
cellulose, i, iii, viii, 5, 7, 15, 16, 18, 20, 21, 23, 24, 25, 26, 27, 32, 44, 50, 51, 53, 57, 58, 59, 60
cellulose derivatives, 2
chiral, 27
chitin, 1, 30
chitosan, 1, 30
classical, 7
cleavage, 26, 31
C-N, 40
CO_2, 15
co-existence, 20
cohesion, 19
competition, 32
components, 20, 35
composites, vii, 2, 3
composition, 21, 58
compounds, 28
concentration, 10, 11, 23, 32, 33
concordance, 45
conjugation, vii
connectivity, 40, 41
contact time, 11
conversion, 2, 7, 10, 12, 13, 23, 28, 31, 34, 37, 38, 39, 44, 45, 47, 51, 52
correlation, 39, 41
correlations, 41

cotton, 1, 5, 7, 16, 21, 26, 27, 32, 33, 34, 45, 46, 47, 49, 50, 51
coupling, 40, 41
critical value, 27
crosslinking, 35
crystalline, 1, 2, 7, 18, 19, 20, 21, 24, 25, 26, 37, 48, 50, 51
crystallinity, 2, 11, 19, 21, 23, 44, 48, 50
crystallisation, 23
crystallites, 50, 52
crystallization, 23
crystals, 2, 7, 25, 52, 53
cuticle, 16
cutin, 16

D

decomposition, 29
decompression, 7
deconvolution, 11, 50
decoupling, 11
defects, 25, 26
degradation, vii, 28, 31, 32, 33, 45
degree of crystallinity, 11, 20
density, 20, 34
derivatives, 2, 7, 12, 35, 37, 47, 60
dialysis, 2, 8, 45
diffraction, 20, 21, 23, 24
dimerization, 29
dimorphism, 21
dislocations, 25
disorder, 50
dispersion, 8
displacement, 35, 36, 52
distilled water, 8
distribution, 37
donor, 29
drying, 45

E

earth science, 61
electron, viii, 21, 23, 24, 29
electron diffraction, 21, 24

electron microscopy, 24
electron paramagnetic resonance, viii
Electron Paramagnetic Resonance, 11, 47
energy, 29
environment, 48, 50
enzymatic, 24, 28
epidermis, 16
EPR, viii, 2, 11, 12, 37, 38, 45, 47, 55, 56
equilibrium, 17
ester, 35, 36, 39, 55, 56
esterification, 2
esters, 27
ethanol, 8, 9
evolution, 32

F

family, 16, 24
fiber, 15, 16, 21, 26
fibers, 24
fibrillar, 24
filtration, 9
financial support, 56
fixation, 15
folding, 25
fourier, vii
fragmentation, 23
free radical, 47
freeze-dried, 8, 9
FTIR, 2, 21, 34, 35, 37, 38, 39, 45, 46, 47, 55
FT-IR, viii
FT-IR, 10
FTIR spectroscopy, 39, 45
fungi, vii

G

gas, 15
gelation, 2
generation, 30
glucose, 17, 18

glycerol, 21
glycol, 31
glycosyl, 6, 8, 12, 13, 33, 37, 50
goals, viii, 2
government, iv
grafting, 1, 2, 6, 8, 10, 26, 35, 36, 37, 39, 40, 41, 44, 45, 47, 48, 50, 51, 52, 53, 55, 56
grafting reaction, 6, 8, 10, 26, 35, 36, 37, 39, 44, 45, 50, 52, 53, 55, 56
groups, vii, viii, 2, 9, 10, 17, 18, 27, 28, 29, 31, 32, 33, 34, 35, 37, 40, 44, 47, 48, 50, 52, 55, 56
growth, 15

H

H1, 40, 41, 43
H_2, 40, 43
hemicellulose, 15
high pressure, 7
high resolution, 48
high temperature, 21
hydrochloric acid, 9, 27
hydrogen, 2, 7, 10, 18, 19, 23, 24, 34, 45, 52
hydrogen bonds, 10, 18, 19, 23, 24, 34, 45, 52
hydrolysis, 1, 2, 7, 26, 27, 28, 35, 51, 52, 55, 59
hydrolyzed, 1, 7, 26, 33, 50, 55
hydroxide, 32
hydroxyl, 1, 17, 18, 27, 28, 29, 31, 32, 33, 34, 35, 37, 44
hydroxyl groups, 1, 17, 18, 27, 28, 31, 32, 33, 34, 37, 44

I

images, 13
in situ, 1, 30
indication, 47, 50
individualization, 52
industrial, 26

inert, 30
inertness, 29
infrared, 20
infrared spectroscopy, 20
instability, 29
integration, 11, 47, 48, 50
integrity, 2, 7, 50
interface, 25
intermolecular, 24
inulin, 1, 30
ions, 39
IR, viii, 10

K

KBr, 10

L

lattice, 7, 51
lignin, 15
limitation, 44
linear, viii, 2, 16, 37, 39, 55
linkage, viii, 17

M

magnetic, iv, 11
magnetic field, 11
MAS, 11, 48, 49, 50, 51
matrix, 15
measurement, 45
media, 1, 2, 8
medicine, vii
methanol, 24
methyl group, 29
methyl groups, 29
microscopy, 23, 24
microwave, 11
migration, 36
milligrams, 10
mixing, 35
models, 51
modulation, 11

moieties, viii, 35
molar ratio, 50
molecular weight, 12, 28, 31, 56
molecules, 1, 2, 7, 19, 29, 44, 45, 55
monosaccharides, 28
morphological, 23
morphology, 55
motion, 47
movement, 15
mushrooms, 21

N

nanocomposites, vii
nanocrystal, 8, 33, 53, 55
nanocrystals, viii, 1, 2, 8, 9, 11, 12, 26, 27, 31, 32, 33, 34, 44, 45, 46, 47, 48, 49, 50, 51, 52, 53, 55, 56
natural, vii, 20, 28
nematic, 27
network, 2, 15, 34, 52
NHS, 5, 6, 8, 35, 36, 37, 39
nitrate, 28
nitric acid, 28
nitrogen, 13, 28, 29
nitrogen dioxide, 28
nitroxide, 28, 29, 41, 45, 47
nitroxide radicals, 28, 29
NMR, viii, 2, 10, 11, 21, 23, 24, 34, 37, 39, 40, 41, 44, 45, 48, 49, 50, 51, 55, 56
NO, 29
non-crystalline, 26
nucleic acid, 35
nucleophiles, 35

O

observations, 33, 37, 44, 46, 47, 55
organic, 15, 16, 27, 29, 30, 35
organic solvent, 27
organic solvents, 27
orientation, 17, 19, 24
oxidants, 29, 30

electron microscopy, 24
electron paramagnetic resonance, viii
Electron Paramagnetic Resonance, 11, 47
energy, 29
environment, 48, 50
enzymatic, 24, 28
epidermis, 16
EPR, viii, 2, 11, 12, 37, 38, 45, 47, 55, 56
equilibrium, 17
ester, 35, 36, 39, 55, 56
esterification, 2
esters, 27
ethanol, 8, 9
evolution, 32

F

family, 16, 24
fiber, 15, 16, 21, 26
fibers, 24
fibrillar, 24
filtration, 9
financial support, 56
fixation, 15
folding, 25
fourier, vii
fragmentation, 23
free radical, 47
freeze-dried, 8, 9
FTIR, 2, 21, 34, 35, 37, 38, 39, 45, 46, 47, 55
FT-IR, viii
FT-IR, 10
FTIR spectroscopy, 39, 45
fungi, vii

G

gas, 15
gelation, 2
generation, 30
glucose, 17, 18

glycerol, 21
glycol, 31
glycosyl, 6, 8, 12, 13, 33, 37, 50
goals, viii, 2
government, iv
grafting, 1, 2, 6, 8, 10, 26, 35, 36, 37, 39, 40, 41, 44, 45, 47, 48, 50, 51, 52, 53, 55, 56
grafting reaction, 6, 8, 10, 26, 35, 36, 37, 39, 44, 45, 50, 52, 53, 55, 56
groups, vii, viii, 2, 9, 10, 17, 18, 27, 28, 29, 31, 32, 33, 34, 35, 37, 40, 44, 47, 48, 50, 52, 55, 56
growth, 15

H

H1, 40, 41, 43
H_2, 40, 43
hemicellulose, 15
high pressure, 7
high resolution, 48
high temperature, 21
hydrochloric acid, 9, 27
hydrogen, 2, 7, 10, 18, 19, 23, 24, 34, 45, 52
hydrogen bonds, 10, 18, 19, 23, 24, 34, 45, 52
hydrolysis, 1, 2, 7, 26, 27, 28, 35, 51, 52, 55, 59
hydrolyzed, 1, 7, 26, 33, 50, 55
hydroxide, 32
hydroxyl, 1, 17, 18, 27, 28, 29, 31, 32, 33, 34, 35, 37, 44
hydroxyl groups, 1, 17, 18, 27, 28, 31, 32, 33, 34, 37, 44

I

images, 13
in situ, 1, 30
indication, 47, 50
individualization, 52
industrial, 26

inert, 30
inertness, 29
infrared, 20
infrared spectroscopy, 20
instability, 29
integration, 11, 47, 48, 50
integrity, 2, 7, 50
interface, 25
intermolecular, 24
inulin, 1, 30
ions, 39
IR, viii, 10

K

KBr, 10

L

lattice, 7, 51
lignin, 15
limitation, 44
linear, viii, 2, 16, 37, 39, 55
linkage, viii, 17

M

magnetic, iv, 11
magnetic field, 11
MAS, 11, 48, 49, 50, 51
matrix, 15
measurement, 45
media, 1, 2, 8
medicine, vii
methanol, 24
methyl group, 29
methyl groups, 29
microscopy, 23, 24
microwave, 11
migration, 36
milligrams, 10
mixing, 35
models, 51
modulation, 11

moieties, viii, 35
molar ratio, 50
molecular weight, 12, 28, 31, 56
molecules, 1, 2, 7, 19, 29, 44, 45, 55
monosaccharides, 28
morphological, 23
morphology, 55
motion, 47
movement, 15
mushrooms, 21

N

nanocomposites, vii
nanocrystal, 8, 33, 53, 55
nanocrystals, viii, 1, 2, 8, 9, 11, 12, 26, 27, 31, 32, 33, 34, 44, 45, 46, 47, 48, 49, 50, 51, 52, 53, 55, 56
natural, vii, 20, 28
nematic, 27
network, 2, 15, 34, 52
NHS, 5, 6, 8, 35, 36, 37, 39
nitrate, 28
nitric acid, 28
nitrogen, 13, 28, 29
nitrogen dioxide, 28
nitroxide, 28, 29, 41, 45, 47
nitroxide radicals, 28, 29
NMR, viii, 2, 10, 11, 21, 23, 24, 34, 37, 39, 40, 41, 44, 45, 48, 49, 50, 51, 55, 56
NO, 29
non-crystalline, 26
nucleic acid, 35
nucleophiles, 35

O

observations, 33, 37, 44, 46, 47, 55
organic, 15, 16, 27, 29, 30, 35
organic solvent, 27
organic solvents, 27
orientation, 17, 19, 24
oxidants, 29, 30

Index

oxidation, vii, viii, 1, 5, 8, 9, 11, 27, 28, 29, 30, 31, 32, 33, 35, 37, 45, 48, 51, 52, 55
oxidative, 27
oxide, 28
oxygen, 19, 29

P

PA, 57
paramagnetic, viii
parenchyma, 1, 16, 26
particles, 26
pectin, 15
peptide, 35
permeability, 16
pH, 7, 8, 30, 31, 32, 35, 36, 37, 39, 44, 56
plants, vii, 15, 16, 21
polar groups, 2
polarity, 17, 24
polarization, 11
polymer, vii, 3, 15, 18, 32
polymers, vii
polysaccharide, 28
polysaccharides, vii, viii, 1, 27, 28, 30, 31, 32, 35
poor, 2, 44
power, 11
precipitation, 45
prediction, 21
pressure, 7
production, 52
proteins, 15, 35
protocol, 7, 24
protons, 40
pulp, 16, 26, 27
pulses, 11
purification, 45

R

radiation, 13
range, 2, 25, 36, 39, 41

raw material, 48
raw materials, 48
reactants, 33
reaction rate, 28
reactive groups, 27
reactivity, viii, 2, 7, 24, 37, 39, 44, 51, 56
reagent, 2, 31, 35
reagents, 2, 5, 6, 27, 29, 35
recalling, 15
redox, 29
regenerate, 31
regenerated cellulose, 20
regeneration, 20, 30, 31
regular, 19
relationships, 20
relaxation, 11
research and development, 26
residues, 17, 18, 24
resistance, 2, 15
resolution, 10, 48
rigidity, 19
room temperature, 11

S

salt, 10, 13, 35
salts, 39
sample, 9, 10, 11, 12, 13, 37, 47
scattering, 24
Schmid, 61
search, viii, 26
secrete, 15
seed, 26
selectivity, 27, 28, 29, 32, 50
series, 11
shape, 1, 11, 26
signals, 39, 40, 41, 48, 55
sites, vii
sodium, 5, 10, 13, 27, 28, 30, 31, 32
sodium hydroxide, 32
software, 13
solid state, viii, 24
solid-state, 2, 7, 21, 23, 34, 48
solubility, 11, 37, 51, 55

solvent, 11
solvents, 27
spatial, 48
species, 15, 47
spectroscopy, viii, 20, 21, 39, 45, 47, 48
spectrum, 26, 39, 40, 43, 47, 48
speed, 8, 11
spin, 11, 47
stability, 23, 29
starch, 1, 28, 30
steric, 32, 37, 44, 51, 53
stretching, 39, 45, 46, 56
structural defect, 26
structural defects, 26
structural knowledge, 26
substrates, 8, 12, 29, 30, 44, 52, 55, 56
sugar, 16, 26, 27
sugar beet, 16, 26, 27
sulfate, 27
sulfuric acid, 27
supernatant, 8
superposition, 10
surface area, 48
suspensions, 8, 27
swelling, 7, 23, 24
symmetry, 24
synchrotron, 21
synthesis, 35
systems, viii, 3, 35

T

technology, 7, 60
temperature, 11, 21, 23, 27, 32, 56
tensile, 24
tension, 15
testes, 24
textile, vii, 2
three-dimensional, 17, 20
time, 9, 11, 23, 24, 27
tissue, 58

titration, viii, 9, 32
transfer, 11, 29
transformation, 26
transition, 20, 23
transmission, 24
transmission electron microscopy, 24
turgor, 15
two step method, 31

U

urea, 35

V

vacuum, 13
values, 33, 36, 39, 45, 51, 52, 55
variation, 8, 34
vibration, 39, 45, 46, 47, 56

W

water, 1, 2, 7, 8, 10, 23, 24, 27, 30, 31, 34, 35, 37, 44, 45, 51
water-soluble, viii, 1, 2, 8, 30, 31, 34, 35, 45
WAXS, 24
wood, 16, 27

X

X-ray diffraction, 20, 21, 23
X-rays, 21

Y

yield, 1, 8, 28, 31, 33, 34